HOW FORESTS THINK

HOW FORESTS THINK

TOWARD AN ANTHROPOLOGY
BEYOND THE HUMAN

Eduardo Kohn

UNIVERSITY OF CALIFORNIA PRESS

Berkeley   Los Angeles   London

University of California Press, one of the most distinguished
university presses in the United States, enriches lives around the
world by advancing scholarship in the humanities, social sciences,
and natural sciences. Its activities are supported by the UC Press
Foundation and by philanthropic contributions from individuals
and institutions. For more information, visit www.ucpress.edu.

University of California Press
Berkeley and Los Angeles, California

University of California Press, Ltd.
London, England

Library of Congress Cataloging-in-Publication Data

Kohn, Eduardo.
  How forests think : toward an anthropology beyond the
human / Eduardo Kohn.
    p. cm.
  Includes bibliographical references and index.
  ISBN 978-0-520-27610-9 (cloth : alk. paper)
  ISBN 978-0-520-27611-6 (pbk. : alk. paper)
    1. Quichua Indians.   2. Quechua Indians—Social life
and customs.   3. Quechua mythology.   4. Indigenous
peoples—Ecology—Amazon River Region.   5. Human-
animal relationships—Amazon River Region.   6. Human-
plant relationships—Amazon River Region.   7. Philosophy
of nature—Amazon River Region.   8. Semiotics—Amazon
River Region.   9. Social sciences—Amazon River Region—
Philosophy.   I. Title
  F2230.2.K4+
  986.6—dc23   2013003750

Manufactured in the United States of America

22 21 20 19 18 17 16
10 9 8

In keeping with a commitment to support environmentally
responsible and sustainable printing practices, UC Press has
printed this book on Natures Natural, a fiber that contains 30%
post-consumer waste and meets the minimum requirements of
ANSI/NISO Z39.48–1992 (R 1997) (Permanence of Paper).

In memory of my grandmother Costanza Di Capua, who, borrowing her words from Gabriele D'Annunzio, would say to me

Io ho quel che ho donato

[I have what I have given]

And for Lisa, who helps me learn how to give this gift

CONTENTS

ACKNOWLEDGMENTS

How Forests Think has been gestating for some time, and I have many to thank
for the life it has taken. I am indebted foremost to the people of Ávila. The
times I spent in Ávila have been some of the happiest, most stimulating, and
also most tranquil I have known. I hope that the sylvan thinking I learned to
recognize there can continue to grow through this book. *Pagarachu.*

Before I even went to Ávila, my grandparents the late Alberto and Costanza
Di Capua had already prepared the way. Italian Jewish refugees settled in
Quito, they brought their curiosity to everything around them. In the 1940s
and 1950s my grandfather, a pharmaceutical chemist, participated in several
scientific expeditions to the Amazon forests in search of plant remedies. My
grandmother, a student of art history and literature in Rome, the city of her
birth, turned to archaeology and anthropology in Quito as a way of understanding better the world into which she had been thrown and which she
would eventually call home. Nonetheless, when I returned from my trips to
Ávila she would insist I read to her from Dante's *Divine Comedy* while she
finished her evening soup. Literature and anthropology were never far removed
for her or for me.

I was twelve years old when I met Frank Salomon in my grandmother's
study. Salomon, a scholar like no other, and the person who would eventually
direct my PhD research at Wisconsin, taught me to see poetry as ethnography
by other means and so opened the space for writing about things as strange
and real as thinking forests and dreaming dogs. The University of Wisconsin–
Madison was a wonderful environment for thinking about the Upper Amazon

in its cultural, historical, and ecological contexts. I owe a great debt also to Carmen Chuquín, Bill Denevan, Hugh Iltis, Joe McCann, Steve Stern, and Karl Zimmerer.

I had the good fortune to write my dissertation—the first stab at what I'm trying to do in this book—at the School for Advanced Research in Santa Fe, thanks to a Weatherhead Resident Scholar Fellowship. There I am indebted to James Brooks, Nancy Owen Lewis, and Doug Schwartz. I am also grateful to the other resident scholars in my cohort: Brian Klopotek, David Nugent, Steve Plog, Barbara Tedlock and Dennis Tedlock, and especially Katie Stewart, who was always ready to talk about ideas as we hiked through the Santa Fe hills.

It was as a Woodrow Wilson Postdoctoral Fellow at the Townsend Center for the Humanities, Berkeley, that I began to develop the conceptual framework for thinking anthropologically beyond the human. I wish especially to acknowledge Candace Slater, as well as Tom Laqueur and Louise Fortmann, for this opportunity. I am also most grateful to my anthropology mentors at Berkeley. Bill Hanks made me part of the anthropological community and sagely guided me, Lawrence Cohen believed in me even when I didn't, and Terry Deacon, in large part through his "pirates" seminar (with participants Ty Cashman, James Haag, Julie Hui, Jay Ogilvy, and Jeremy Sherman), created the most intellectually stimulating environment I have ever been in and forever changed the way I think. Four friends and colleagues from those Berkeley days deserve a special mention: Liz Roberts, who taught me so much about anthropology (and who also introduced me to all the right people), Cristiana Giordano, Pete Skafish, and Alexei Yurchak. The members of the anthropology department were extremely kind and supportive. Thanks especially to Stanley Brandes, Meg Conkey, Mariane Ferme, Rosemary Joyce, Nelson Graburn, Christine Hastorf, Cori Hayden, Charles Hirschkind, Don Moore, Stefania Pandolfo, Paul Rabinow, and Nancy Scheper-Hughes.

At the Michigan Society of Fellows, I wish to thank the former director Jim White and the fellows, especially Paul Fine, Stella Nair, Neil Safier, and Daniel Stolzenberg, with whom I spent two wonderful years. At the University of Michigan's Department of Anthropology, I am indebted to Ruth Behar, the late Fernando Coronil, Webb Keane, Stuart Kirsch, Conrad Kottak, Alaina Lemon, Bruce Mannheim, Jennifer Robertson, Gayle Rubin, Julie Skurski, and Katherine Verdery, as well as the members of my writing group, Rebecca Hardin, Nadine Naber, Julia Paley, Damani Partridge, and Miriam Ticktin.

I also wish to express gratitude to my former Cornell colleagues, especially to Stacey Langwick, Michael Ralph, Nerissa Russell, Terry Turner, Marina Welker, Andrew Wilford, and, above all, Hiro Miyazaki and Annelise Riles, who generously organized (with the participation of Tim Choy, Tony Crook, Adam Reed, and Audra Simpson) a workshop on my book manuscript.

In Montreal I have found a stimulating place to think, teach, and live. My colleagues at McGill have supported me in countless ways. I especially wish to thank the following people for reading portions of the manuscript and/or discussing parts of the project: Colin Chapman, Oliver Coomes, Nicole Couture, John Galaty, Nick King, Katherine Lemons, Margaret Lock, Ron Niezen, Eugene Raikhel, Tobias Rees, Alberto Sánchez, Colin Scott, George Wenzel, and Allan Young. Thank you also to my wonderful undergraduates, especially those who took the courses "Anthropology and the Animal" and "Anthropology beyond the Human." I am also grateful to the graduate students who read and critically engaged parts of my book manuscript: Amy Barnes, Mónica Cuéllar, Darcie De Angelo, Arwen Fleming, Margaux Kristjansson, Sophie Llewelyn, Brodie Noga, Shirin Radjavi, and Daniel Ruiz Serna. Finally, I am indebted to Sheehan Moore for the help he provided as my able research assistant.

Many people in Montreal and elsewhere have, over the years, supported and inspired my work. First and foremost I wish to thank Donna Haraway. Her refusal to allow me to grow complacent in my thinking is for me the mark of a true friend. I also wish to thank Pepe Almeida, Angel Alvarado, Felicity Aulino, Gretchen Bakke, Vanessa Barreiro, João Biehl, Michael Brown, Karen Bruhns, Matei Candea, Manuela Carneiro da Cunha, Michael Cepek, Chris Chen, John Clark, Biella Coleman, André Costopoulos, Mike Cowan, Veena Das, Nais Dave, Marisol de la Cadena, MaryJo DelVecchio Good, Bob Desjarlais, Nick Dew, Alicia Díaz, Arcadio Díaz Quiñones, Didier Fassin, Carlos Fausto, Steve Feld, Allen Feldman, Blenda Femenias, Enrique Fernández, Jennifer Fishman, Agustín Fuentes, Duana Fullwiley, Chris Garces, Fernando García, the late Clifford Geertz, Ilana Gershon, Eric Glassgold, Maurizio Gnerre, Ian Gold, Byron Good, Mark Goodale, Peter Gose, Michel Grignon, Geoconda Guerra, Rob Hamrick, Clara Han, Susan Harding, Stefan Helmreich, Michael Herzfeld, Kregg Hetherington, Frank Hutchins, Sandra Hyde, Tim Ingold, Frédéric Keck, Chris Kelty, Eben Kirksey, Tom Lamarre, Hannah Landecker, Bruno Latour, Jean Lave, Ted Macdonald, Setrag Manoukian, Carmen Martínez, Ken Mills, Josh Moses, Blanca Muratorio, Paul Nadasdy, Kristin Norget, Janis

Nuckolls, Mike Oldani, Ben Orlove, Anand Pandian, Héctor Parión, Morten Pederson, Mario Perín, Michael Puett, Diego Quiroga, Hugh Raffles, Lucinda Ramberg, Charlie Reeves, Lisa Rofel, Mark Rogers, Marshall Sahlins, Fernando Santos-Granero, Patrice Schuch, Natasha Schull, Jim Scott, Glenn Shepard, Kimbra Smith, Barb Smuts, Marilyn Strathern, Tod Swanson, Anne-Christine Taylor, Lucien Taylor, Mike Uzendoski, Ismael Vaccaro, Yomar Verdezoto, Eduardo Viveiros de Castro, Norm Whitten, Eileen Willingham, Yves Winter, and Gladys Yamberla.

Many tropical biologists taught me over the years about their field and allowed me to bounce ideas off of them. David Benzing and Steve Hubbell were early mentors. Thanks also to Selene Baez, Robyn Burnham, Paul Fine, and Nigel Pitman. I am grateful for having had the opportunity to immerse myself in this area of study through the tropical ecology field course run by the Organization for Tropical Studies (OTS) in Costa Rica. Quito has a vibrant and warm community of biologists, and I thank the late Fernando Ortíz Crespo, Giovanni Onore and Lucho Coloma at the Universidad Católica, as well as Walter Palacios, Homero Vargas, and especially David Neill at the Herbario Nacional del Ecuador for so generously taking me in. This project involves a sizable ethnobiological component, and I am grateful to all the specialists who helped me identify my specimens. I would especially like to thank David Neill once again for his careful revision of my botanical collections. I am also indebted to Efraín Freire for his work with these collections. For their botanical determinations, I wish to acknowledge the following individuals (followed by the herbaria they were affiliated with when they made the identifications): M. Asanza (QCNE), S. Baez (QCA), J. Clark (US), C. Dodson (MO), E. Freire (QCNE), J. P. Hedin (MO), W. Nee (NY), D. Neill (MO), W. Palacios (QCNE), and T. D. Pennington (K). I wish to thank G. Onore, as well as M. Ayala, E. Baus, C. Carpio, all at the time at QCAZ; and D. Roubick (STRI) for determining my invertebrate collections. I wish to thank L. Coloma as well as J. Guayasamín and S. Ron, at the time at QCAZ, for determining my herpetofauna collections. I am grateful to P. Jarrín (QCAZ) for determining my mammal collections. Finally, I wish to thank Ramiro Barriga of the Escuela Politécnica Nacional for determining my fish collections.

This project would not have been possible without the generous support of many institutions. I am grateful for a Fulbright Grant for Graduate Study and Research Abroad, a National Science Foundation Graduate Fellowship, a Fulbright-Hays Doctoral Research Abroad Grant, a University of

Wisconsin–Madison Latin American and Iberian Studies Field Research Grant, a Wenner-Gren Foundation for Anthropological Research Pre-Doctoral Grant, and a grant from the Fonds québécois de la recherche sur la société et la culture (FQRSC).

I was lucky enough to have had the opportunity to present the book's entire argument through visiting professorships at Oberlin College (for which I thank Jack Glazier) and at L'École des hautes études en sciences sociales (EHESS) by the generous invitation of Philippe Descola. I have also presented portions of the argument at Carleton University, the University of Chicago, the Facultad Latinoamericana de Ciencias Sociales sede Ecuador (FLACSO), Johns Hopkins University, the University of California, Los Angeles, the University of California, Santa Cruz, the University of Toronto, and Yale University. An earlier version of chapter 4 appeared in *American Ethnologist*.

Numerous people have engaged with the book as a whole. I cannot thank Olga González, Josh Reno, Candace Slater, Anna Tsing, and Mary Weismantel enough for their stimulating, thoughtful, and constructive reviews. I am grateful to David Brent, Priya Nelson, and Jason Weidemann for their sustained interest in this project. I wish to give special thanks to Pete Skafish and Alexei Yurchak, who took time from their busy lives to carefully read large portions of the book (and to discuss them with me at length via Skype), and I am especially indebted to Lisa Stevenson for critically reading and meticulously editing the entire manuscript. Finally, I wish to thank Reed Malcolm, my editor at the University of California Press, for being jazzed by what must surely have seemed like a risky project. I also wish to thank Stacy Eisenstark; my patient copy editor, Sheila Berg; and my project manager, Kate Hoffman.

I owe so much to my family members for all they have given me. I could not have had a more generous uncle than Alejandro Di Capua. I wish to thank him and his family for always welcoming me into their Quito home. My uncle Marco Di Capua, who shares my love for Latin American history and science, was, along with his family, always interested in hearing about my work, and for this I am most grateful. I also wish to express my gratitude to Riccardo Di Capua and all my Ecuadorian Kohn cousins. I especially wish to thank the late Vera Kohn for reminding me how to think in wholes.

I am fortunate to have had the unfailing love and support of my parents, Anna Rosa and Joe, and my sisters, Emma and Alicia. My mother was the first to teach me to notice things in the woods; my father, how to think for myself; and my sisters, how to think about others.

I am indebted to my mother-in-law, Frances Stevenson, who spent several of her summer vacations on lakes in Quebec, Ontario, and the Adirondacks watching the kids while I wrote. I am also grateful to my father-in-law, Romeyn Stevenson, and his wife, Christine, for understanding that this other kind of "work" I always seemed to bring to the farm would keep me away from many more pressing chores.

Finally, thank you, Benjamin and Milo, for putting up with all this "'versity" stuff, as you put it. You teach me every day how to see my university work as your kind of play. *Gracias*. And thank you, Lisa, for everything; for inspiring me, for helping me both to grow and to recognize my limits, and for being such a wonderful companion in this life of ours.

Introduction: *Runa Puma*

*Ahi quanto a dir qual era è cosa dura
esta selva selvaggia e aspra e forte . . .*

[Ah, it is hard to speak of what it was
that savage forest, dense and difficult . . .]

—Dante Alighieri, *The Divine Comedy*, *Inferno*, Canto I [trans. Mandelbaum]

Settling down to sleep under our hunting camp's thatch lean-to in the foothills of Sumaco Volcano, Juanicu warned me, "Sleep faceup! If a jaguar comes he'll see you can look back at him and he won't bother you. If you sleep facedown he'll think you're *aicha* [prey; lit., "meat" in Quichua] and he'll attack." If, Juanicu was saying, a jaguar sees you as a being capable of looking back—a self like himself, a *you*—he'll leave you alone. But if he should come to see you as prey—an *it*—you may well become dead meat.[1]

How other kinds of beings see us matters. That other kinds of beings see us changes things. If jaguars also represent us—in ways that can matter vitally to us—then anthropology cannot limit itself just to exploring how people from different societies might happen to represent them as doing so. Such encounters with other kinds of beings force us to recognize the fact that seeing, representing, and perhaps knowing, even thinking, are not exclusively human affairs.

How would coming to terms with this realization change our understandings of society, culture, and indeed the sort of world that we inhabit? How does it change the methods, scope, practice, and stakes of anthropology? And, more important, how does it change our understanding of anthropology's object—the "human"—given that in that world beyond the human we sometimes find things we feel more comfortable attributing only to ourselves?

That jaguars represent the world does not mean that they necessarily do so as we do. And this too changes our understanding of the human. In that realm beyond the human, processes, such as representation, that we once thought we understood so well, that once seemed so familiar, suddenly begin to appear strange.

So as not to become meat we must return the jaguar's gaze. But in this encounter we do not remain unchanged. We become something new, a new kind of *"we"* perhaps, aligned somehow with that predator who regards us as a predator and not, fortunately, as dead meat. The forests around Juanicu's Quichua-speaking Runa village, Ávila, in Ecuador's Upper Amazon (a village that is a long day's hike from that makeshift shelter under which we, that night, were diligently sleeping faceup) are haunted by such encounters.[2] They are full of *runa puma*, shape-shifting human-jaguars, or were-jaguars as I will call them.

Runa in Quichua means "person"; *puma* means "predator" or "jaguar." These runa puma—beings who can see themselves being seen by jaguars as fellow predators, and who also sometimes see other humans the way jaguars do, namely, as prey—have been known to wander all the way down to the distant Napo River. The shamans in Río Blanco, a Runa settlement on the banks of the Upper Napo where I worked in the late 1980s, would see these were-jaguars in their *aya huasca*-induced visions.[3] "The runa puma that walk the forests around here," one shaman told me, "they're from Ávila." They described these massive runa puma as having white hides. The Ávila Runa, they insisted, become jaguars, white were-jaguars, *yura runa puma*.

Ávila enjoys a certain reputation in the Runa communities of the Upper Napo. "Be careful going up to Ávila," I was cautioned. "Be especially wary of their drinking parties. When you go out to pee you might come back to find that your hosts have become jaguars." In the early 1990s, in Tena, the capital of Napo Province, a friend and I went out drinking one night at a *cantina*, a makeshift tavern, with some of the leaders of FOIN, the provincial indigenous federation. Amid boasts of their own prowess—Who could command the most support from the base communities? Who could best bring in the big NGO checks?—talk turned more specifically to shamanic power and where the seat of such power, the font of FOIN's strength, really lay. Was it, as some that night held, Arajuno, south of the Napo? This is an area of Runa settlement that borders on the east and south with the Huaorani, a group that many Runa view with a mixture of fear, awe, and disdain as "savage" (*auca* in Quichua, hence their pejorative ethnonym Auca). Or was it Ávila, home to so many runa puma?

That night around the cantina table Ávila edged out Arajuno as a center of power. This village at first might seem an unlikely choice to signify shamanic power in the figure of a jaguar. Its inhabitants, as they would be the first to insist, are anything but "wild." They are, and, as they invariably make clear, have always been Runa—literally, "human persons"—which for them means that they have always been Christian and "civilized." One might even say that they are, in important but complicated ways (ways explored in the final chapter), "white." But they are, some of them, also equally—and really—puma.[4]

Ávila's position as a seat of shamanic power derives not just from its relation to some sort of sylvan savagery but also from its particular position in a long colonial history (see figure 1). Ávila was one of the earliest sites of Catholic indoctrination and Spanish colonization in the Upper Amazon. It was also the epicenter of a late-sixteenth-century regionally coordinated uprising against the Spaniards.

That rebellion against the Spaniards, a response in part to the increasingly onerous burden of tribute payment, was, according to colonial sources, sparked by the visions of two shamans. Beto, from the Archidona region, saw a cow who "spoke with him . . . and told him that the God of the Christians was very angry with the Spaniards who were in that land." Guami, from the Ávila region, was "transported out of this life for five days during which he saw magnificent things, and the God of the Christians sent him to kill everyone and burn their houses and crops" (de Ortiguera 1989 [1581–85]: 361).[5] In the uprising that ensued the Indians around Ávila did, according to these sources, kill all the Spaniards (save one, about whom more in chapter 3), destroy their houses, and eradicate the orange and fig trees and all the other foreign crops from the land.

These contradictions—that Runa shamans receive messages from Christian gods and that the were-jaguars that wander the forests around Ávila are white—are part of what drew me to Ávila. The Ávila Runa are far removed from any image of a pristine or wild Amazon. Their world—their very being— is thoroughly informed by a long and layered colonial history. And today their village is just a few kilometers from the growing, bustling colonist town of Loreto and the expanding network of roads that connects this town with increasing efficiency to the rest of Ecuador. And yet they also live intimately with all kinds of real jaguars that walk the forests around Ávila; these include those that are white, those that are Runa, and those that are decidedly spotted.

This intimacy in large part involves eating and also the real risk of being eaten. A jaguar killed a child when I was in Ávila. (He was the son of the

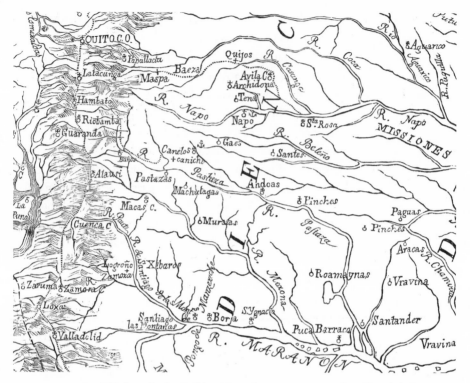

FIGURE 1. As visible from the detail of the eighteenth-century map reproduced here (which corresponds very roughly to modern Ecuador's Andean and Amazonian regions), Ávila (upper center) was considered a missionary center (represented by a cross). It was connected by foot trails (dotted line) to other such centers, such as Archidona, as well as to the navigable Napo River (a tributary of the Amazon), and to Quito (upper left). The linear distance between Quito and Ávila is approximately 130 kilometers. The map indicates some of the historical legacies of colonial networks in which Ávila is immersed; the landscape of course has not remained unchanged. Loreto, the major colonist town, approximately 25 kilometers east of Ávila, is wholly absent from the map, though it figures prominently in the lives of the Ávila Runa and in this book. From Requena 1779 [1903]. Collection of the author.

woman posing with her daughter in the photograph that serves as the frontis-piece for this chapter, a photograph the mother asked me to take so that she might have some memory of her daughter if she too were taken away.) And jaguars, as I discuss later in this book, also killed several dogs during my time in Ávila. They also shared their food with us. On several occasions we found half-eaten carcasses of agoutis and pacas that were-jaguars had left for us in the forest as gifts and that subsequently became our meals. Felines of all kinds, including these generous meat-bearing runa puma, are sometimes hunted.

Eating also brings people in intimate relation to the many other kinds of nonhuman beings that make the forest their home. During the four years that I worked in Ávila villagers bought many things in Loreto. They bought things such as shotguns, ammunition, clothing, salt, many of the household items that would have been made by hand a couple of generations ago, and lots of the contraband cane liquor that they call *cachihua*. What they didn't buy was food. Almost all the food they shared with each other and with me came from their gardens, the nearby rivers and streams, and the forest. Getting food through hunting, fishing, gathering, gardening, and the management of a variety of ecological assemblages involves people intimately with one of the most complex ecosystems in the world—one that is chock-full of an astounding array of different kinds of interacting and mutually constituting beings. And it brings them into very close contact with the myriad creatures—and not just jaguars—that make their lives there. This involvement draws people into the lives of the forest. It also entangles the lives of that forest with worlds we might otherwise consider "all too human," by which I mean the moral worlds we humans create, which permeate our lives and so deeply affect those of others.

Gods talking through the bodies of cows, Indians in the bodies of jaguars, jaguars in the clothing of whites, the runa puma enfolds these. What are we anthropologists—versed as we are in the ethnographic charting of the distinctive meaning-filled morally loaded worlds we humans create (distinctive worlds that make us feel that we are exceptions in this universe)—to make of this strange other-than-human and yet all-too-human creature? How should we approach this Amazonian Sphinx?

Making sense of this creature poses a challenge not unlike the one posed by that other Sphinx, the one Oedipus encountered on his way to Thebes. That Sphinx asked Oedipus, "What goes on four legs in the morning, on two legs at noon, and on three legs in the evening?" To survive this encounter Oedipus, like the members of our hunting party, had to figure out how to correctly respond. His answer to the riddle the Sphinx posed from her position somewhere (slightly) beyond the human was, "Man." It is a response that, in light of the Sphinx's question, begs us to ask, What are we?

That other-than-human Sphinx whom, despite her inhumanity, we nevertheless regard and to whom we must respond, asks us to question what we think we know about the human. And her question reveals something about our answer. Asking what first goes on four, then on two, then on three legs simultaneously invokes the shared legacies of our four-pawed animality and our distinctively bipedal peripatetic humanity, as well the various kinds of canes we fashion and incorporate to feel our ways through our finite lives—lives whose ends, as Kaja Silverman (2009) observes, ultimately connect us to all the other beings with whom we share the fact of finitude.

Footing for the unsteady, a guide for the blind, a cane mediates between a fragile mortal self and the world that spans beyond. In doing so it represents something of that world, in some way or another, to that self. Insofar as they serve to represent something of the world to someone, many entities exist that can function as canes for many kinds of selves. Not all these entities are artifacts. Nor are all these kinds of selves human. In fact, along with finitude, what we share with jaguars and other living selves—whether bacterial, floral, fungal, or animal—is the fact that how we represent the world around us is in some way or another constitutive of our being.

A cane also prompts us to ask with Gregory Bateson, "where" exactly, along its sturdy length, "do *I* start?" (Bateson 2000a: 465). And in thus highlighting representation's contradictory nature—Self or world? Thing or thought? Human or not?—it indicates how pondering the Sphinx's question might help us arrive at a more capacious understanding of Oedipus's answer.

This book is an attempt to ponder the Sphinx's riddle by attending ethnographically to a series of Amazonian other-than-human encounters. Attending to our relations with those beings that exist in some way beyond the human forces us to question our tidy answers about the human. The goal here is neither to do away with the human nor to reinscribe it but to open it. In rethinking the human we must also rethink the kind of anthropology that would be adequate to this task. Sociocultural anthropology in its various forms as it is practiced today takes those attributes that are distinctive to humans—language, culture, society, and history—and uses them to fashion the tools to understand humans. In this process the analytical object becomes isomorphic with the analytics. As a result we are not able to see the myriad ways in which people are connected to a broader world of life, or how this fundamental connection changes what it might mean to be human. And this is why expanding ethnography to reach beyond the human is so important. An ethnographic focus not just on humans or only on animals but also on how humans and animals relate breaks open the circular closure that otherwise confines us when we seek to understand the distinctively human by means of that which is distinctive to humans.

Creating an analytical framework that can include humans as well as nonhumans has been a central concern of science and technology studies (see esp.

Latour 1993, 2005), the "multispecies" or animal turn (see esp. Haraway 2008; Mullin and Cassidy 2007; Choy et al. 2009; see also Kirksey and Helmreich 2010 for a review), and Deleuze-influenced (Deleuze and Guattari 1987) scholarship (e.g., Bennett 2010). Along with these approaches I share the fundamental belief that social science's greatest contribution—the recognition and delimitation of a separate domain of socially constructed reality—is also its greatest curse. Along with these I also feel that finding ways to move beyond this problem is one of the most important challenges facing critical thought today. And I have especially been swayed by Donna Haraway's conviction that there is something about our everyday engagements with other kinds of creatures that can open new kinds of possibilities for relating and understanding.

These "posthumanities" have been remarkably successful at focusing on the zone beyond the human as a space for critique and possibility. However, their productive conceptual engagement with this zone is hampered by certain assumptions, shared with anthropology and social theory more broadly, concerning the nature of representation. Furthermore, in attempting to address some of the difficulties these assumptions about representation create, they tend to arrive at reductionistic solutions that flatten important distinctions between humans and other kinds of beings, as well as those between selves and objects.

In *How Forests Think* I seek to contribute to these posthuman critiques of the ways in which we have treated humans as exceptional—and thus as fundamentally separate from the rest of the world—by developing a more robust analytic for understanding human relations to nonhuman beings. I do so by reflecting on what it might mean to say that forests think. I do so, that is, by working out the connection between representational processes (which form the basis for all thought) and living ones as this is revealed through ethnographic attention to that which lies beyond the human. I use the insights thus gained to rethink our assumptions about the nature of representation, and I then explore how this rethinking changes our anthropological concepts. I call this approach an "anthropology beyond the human."[6]

In this endeavor I draw on the work of the nineteenth-century philosopher Charles Peirce (1931, 1992a, 1998a), especially his work in semiotics (the study of how signs represent things in the world). In particular I invoke what the Chicago-trained linguistic anthropologist Alejandro Paz calls the "weird" Peirce, by which he means those aspects of Peirce's writing that we anthropologists find hard to digest—those parts that reach beyond the human to situate representation in the workings and logics of a broader nonhuman universe out of which we humans come. I also draw greatly on Terrence Deacon's remarkably creative application of Peircean semiotics to biology and to questions of what he calls "emergence" (see Deacon 2006, 2012).

The first step toward understanding how forests think is to discard our received ideas about what it means to represent something. Contrary to our assumptions, representation is actually something more than conventional, linguistic, and symbolic. Inspired and emboldened by Frank Salomon's (2004) pioneering work on the representational logics of Andean knotted cords and Janis Nuckolls's (1996) work on Amazonian sound images, this is an ethnography that explores representational forms that go beyond language. But it does so by going beyond the human. Nonhuman life-forms also represent the world. This more expansive understanding of representation is hard to appreciate because our social theory—whether humanist or posthumanist, structuralist or poststructuralist—conflates representation with language.

We conflate representation with language in the sense that we tend to think of how representation works in terms of our assumptions about how human language works. Because linguistic representation is based on signs that are conventional, systemically related to one another, and "arbitrarily" related to their objects of reference, we tend to assume that all representational processes have these properties. But symbols, those kinds of signs that are based on convention (like the English word *dog*), which are distinctively human representational forms, and whose properties make human language possible, actually emerge from and relate to other modalities of representation. In Peirce's terminology these other modalities (in broad terms) are either "iconic" (involving signs that share likenesses with the things they represent) or "indexical" (involving signs that are in some way affected by or otherwise correlated with those things they represent). In addition to being symbolic creatures we humans share these other semiotic modalities with the rest of nonhuman biological life (Deacon 1997). These nonsymbolic representational modalities pervade the living world—human and nonhuman—and have underexplored properties that are quite distinct from those that make human language special.

Although there are anthropological approaches that do move beyond the symbolic to study the full range of Peircean signs, they locate such signs exclusively inside a human framework. Accordingly, those who use signs are understood to be human, and though signs may be extralinguistic (with the consequence that language can be treated as something more than symbolic) the contexts that make them meaningful are human sociocultural ones (see esp.

Silverstein 1995; Mannheim 1991; Keane 2003; Parmentier 1994; Daniel 1996; on "context," see Duranti and Goodwin 1992).

These approaches fail to recognize that signs also exist well beyond the human (a fact that changes how we should think about human semiosis as well). Life is constitutively semiotic. That is, life is, through and through, the product of sign processes (Bateson 2000c, 2002; Deacon 1997; Hoffmeyer 2008; Kull et al. 2009). What differentiates life from the inanimate physical world is that life-forms represent the world in some way or another, and these representations are intrinsic to their being. What we share with nonhuman living creatures, then, is not our embodiment, as certain strains of phenomenological approaches would hold, but the fact that we all live with and through signs. We all use signs as "canes" that represent parts of the world to us in some way or another. In doing so, signs make us what we are.

Understanding the relationship between distinctively human forms of representation and these other forms is key to finding a way to practice an anthropology that does not radically separate humans from nonhumans. Semiosis (the creation and interpretation of signs) permeates and constitutes the living world, and it is through our partially shared semiotic propensities that multispecies relations are possible, and also analytically comprehensible.

This way of understanding semiosis can help us move beyond a dualistic approach to anthropology, in which humans are portrayed as separate from the worlds they represent, toward a monistic one, in which how humans represent jaguars and how jaguars represent humans can be understood as integral, though not interchangeable, parts of a single, open-ended story. Given the challenges posed by learning to live with the proliferating array of other kinds of life-forms that increasingly surround us—be they pets, weeds, pests, commensals, new pathogens, "wild" animals, or technoscientific "mutants"— developing a precise way to analyze how the human is both distinct from and continuous with that which lies beyond it is both crucial and timely.

This search for a better way to attend to our relations to that which lies beyond the human, especially that part of the world beyond the human that is alive, forces us to make ontological claims—claims, that is, about the nature of reality. That, for example, jaguars in some way or other represent the world demands a general explanation that takes into account certain insights about the way the world is—insights that are garnered from attention to engagements with nonhumans and that are thus not fully circumscribed by any particular human system of understanding them.

As a recent debate makes clear (Venkatesan et al. 2010), ontology, as it circulates in our discipline, is a thorny term. On the one hand, it is often negatively associated with a search for ultimate truths—the kinds that the ethnographic documentation of so many different ways of doing and seeing is so good at debunking (Carrithers 2010: 157). On the other hand, it sometimes seems to function as nothing more than a trendy word for culture, especially when a possessive pronoun precedes it: *our* ontology, say, versus *theirs* (Holbraad 2010: 180).

In mobilizing Amazonian ethnography to think ontologically, I place myself in the company of two eminent anthropologists, Philippe Descola and Eduardo Viveiros de Castro, who have had a great and lasting influence on my research. Their work has gained traction in anthropology because of the ways it renders ontology plural without turning it into culture: different worlds instead of different worldviews (Candea 2010: 175). But the recognition of multiple realities only side steps the question: Can anthropology make general claims about the way the world is?[7] Despite the many problems that making general claims raises—problems that our various forms of relativism struggle to keep at bay—I think anthropology can. And I think anthropology, to be true to the world, must find ways of making such claims, in part because, as I will argue, generality itself is a property of the world and not just something we humans impose on it. And yet, given our assumptions about representation, it seems difficult to make such claims. This book seeks to get beyond this impasse.

I do not, then, wish to enter the ontological from the direction of the human. My goal is not to isolate configurations of ontological propositions that crop up at a particular place or time (Descola 2005). I choose, rather, to enter at a more basic level. And I try to see what we can learn by lingering at that level. I ask, What kinds of insights about the nature of the world become apparent when we attend to certain engagements with parts of that world that reveal some of its different entities, dynamics, and properties?

In sum, an anthropology beyond the human is perforce an ontological one. That is, taking nonhumans seriously makes it impossible to confine our anthropological inquiries to an epistemological concern for how it is that humans, at some particular time or in some particular place, go about making sense of them. As an ontological endeavor this kind of anthropology places us in a special position to rethink the sorts of concepts we use and to develop new ones. In Marilyn Strathern's words, it aims "to create the conditions for new thoughts" (1988: 20).

FIGURE 2. Ávila circa 1992. Photo by author.

Such an endeavor might seem detached from the more mundane worlds of ethnographic experience that serve as the foundations for anthropological argumentation and insight. And yet this project, and the book that attempts to do it justice, is rigorously empirical in the sense that the questions it addresses grow out of many different kinds of experiential encounters that emerged over the course of a long immersion in the field. As I've attempted to cultivate these questions I've come to see them as articulations of general problems that become amplified, and thus made visible, through my struggles to pay ethnographic attention to how people in Ávila relate to different kinds of beings.

This anthropology beyond the human, then, grows out of an intense sustained engagement with a place and those who make their lives there. I have known Ávila, its environs, and the people who live there for a human generation; the infants I was introduced to on my first visit in 1992 were when I last visited in 2010 young parents; their parents are now grandparents, and some of the parents of those new grandparents are now dead (see figure 2). I spent four years (1996–2000) living in Ecuador and conducting fieldwork in Ávila and continue to visit regularly.

The experiential bases for this book are many. Some of the most important encounters with other kinds of beings came on my walks through the forest

FIGURE 3. Drinking beer. Photo by author.

with Runa hunters, others when I was left alone in the forest, sometimes for hours, as these hunters ran off in pursuit of their quarry—quarry that sometimes ended up circling back on me. Still others occurred during my slow strolls at dusk in the forest just beyond the manioc gardens that surround people's houses where I would be privy to the last burst of activity before so many of the forest's creatures settled down for the night.

I spent much of my time trying to listen, often with a tape recorder in hand, to how people in everyday contexts relate their experiences with different kinds of beings. These conversations often took place while drinking manioc beer with relatives and neighbors or while sipping *huayusa* tea around the hearth in the middle of the night (figure 3).[8] The interlocutors here were usually human and usually Runa. But "conversation" also occasionally involved other kinds of beings: the squirrel cuckoo who flew over the house whose call so radically changed the course of discussion down below; the household dogs with whom people sometimes need to make themselves understood; the woolly monkeys and the powerful spirits that inhabit the forest; and even the politicians who trudge up to the village during election season. With all of these, people in Ávila struggle to find channels of communication.

In my pursuit of certain tangibles of the ecological webs in which the Runa are immersed I also compiled many hundreds of ethnobiological specimens. These were identified by specialists, and they are now housed in Ecuador's main herbarium and museums of natural history.[9] Making these collections very quickly gave me some sort of purchase on the forest and its many creatures. It also allowed an entry to people's understandings of ecological relations and gave me a way to articulate this with other bodies of knowledge about the forest world not necessarily bounded by that particular human context. Collecting imposes its own structures on forest relationships, and I was not unaware of the limitations—and motivations—of this search for stable knowledge, as well as the fact that, in some important respects, my efforts as a collector were quite different from Runa ways of engaging with the beings of the forest (see Kohn 2005).

I also sought to pay attention to forest experiences as they resonate through other arenas that are less grounded. Everyday life in Ávila is entangled with that second life of sleep and its dreams. Sleeping in Ávila is not the consolidated, solitary, sensorially deprived endeavor it has so often become for us. Sleep—surrounded by lots of people in open thatch houses with no electricity and largely exposed to the outdoors—is continuously interspersed with wakefulness. One awakens in the middle of the night to sit by the fire and ward off the chill, or to receive a gourd bowl full of steaming huayusa tea, or on hearing the common potoo call during a full moon, or sometimes even the distant hum of a jaguar. And one awakens also to the extemporaneous comments people make throughout the night about those voices they hear. Thanks to these continuous disruptions, dreams spill into wakefulness and wakefulness into dreams in a way that entangles both. Dreams—my own, those of my housemates, the strange ones we shared, and even those of their dogs—came to occupy a great deal of my ethnographic attention, especially because they so often involved the creatures and spirits that people the forest. Dreams too are part of the empirical, and they are a kind of real. They grow out of and work on the world, and learning to be attuned to their special logics and their fragile forms of efficacy helps reveal something about the world beyond the human.

The thinking in this book works itself through images. Some of these come in the form of dreams, but they also appear as examples, anecdotes, riddles, questions, conundrums, uncanny juxtapositions, and even photographs. Such images can work on us if we would let them. My goal here is to create the conditions necessary to make this sort of thinking possible.

This book is an attempt to encounter an encounter, to look back at these looking-backs, to face that which the runa puma asks of us, and to formulate a response. That response is—to adopt a title from one of the books that Peirce never completed (Peirce 1992b)—my "guess at the riddle" that the Sphinx posed. It is my sense of what we can learn when we attend ethnographically to how the Sphinx's question might reconfigure the human. Making claims about and beyond the human in anthropology is dangerous business; we are experts at undermining arguments through appeals to hidden contexts. This is the analytical trump card that every well-trained anthropologist has up her sleeve. In this sense, then, this is an unusual project, and it requires of you, the reader, a modicum of goodwill, patience, and the willingness to struggle to allow the work done here to work itself through you.

This book will not immediately plunge you into the messy entangled, "natural-cultural" worlds (Latour 1993) whose witnessing has come to be the hallmark of anthropological approaches to nonhumans. Rather, it seeks a gentler immersion in a kind of thinking that grows. It begins with very simple matters so that complexity, context, and entanglement can themselves become the objects of ethnographic analysis rather than the unquestioned conditions for it.

As such, the first chapters may seem far removed from an exposition of the complicated, historically situated, power-laden contexts that so deeply inform Runa ways of being—an exposition we justifiably expect from ethnography. But what I am trying to do here matters for politics; the tools that grow from attention to the ways the Runa relate to other kinds of beings can help think possibility and its realization differently. This, I hope, can speak to what Ghassan Hage (2012) calls an "alter-politics"—a politics that grows not from opposition to or critique of our current systems but one that grows from attention to another way of being, one here that involves other kinds of living beings.

This book, then, attempts to develop an analytic, which seeks to take anthropology "beyond the human" but without losing sight of the pressing ways in which we are also "all too human," and how this too bears on living. The first step toward this endeavor, and the subject of the first chapter, "The

Open Whole," is to rethink human language and its relationship to those other forms of representation we share with nonhuman beings. Whether or not it is explicitly stated, language, and its unique properties, is what, according to so much of our social theory, defines us. Social or cultural systems, or even "actor-networks," are ultimately understood in terms of their languagelike properties. Like words, their "relata"—whether roles, ideas, or "actants"—do not precede the mutually constitutive relationships these have with one another in a system that necessarily comes to exhibit a certain circular closure by virtue of this fact.[10]

Given so much of social theory's emphasis on recognizing those unique sorts of languagelike phenomena responsible for such closure, I explore how, thanks to the ways in which language is nested within broader forms of representation that have their own distinctive properties, we are, in fact, open to the emerging worlds around us. In short, if culture is a "complex whole," to quote E. B. Tylor's (1871) foundational definition (a definition that invokes the ways in which cultural ideas and social facts are mutually constituted by virtue of the sociocultural systemic contexts that sustain them), then culture is also an "open whole." The first chapter, then, constitutes a sort of ethnography of signs beyond the human. It undertakes an ethnographic exploration of how humans and nonhumans use signs that are not necessarily symbolic—that is, signs that are not conventional—and demonstrates why these signs cannot be fully circumscribed by the symbolic.

Exploring how such aperture exists despite the very real fact of symbolic closure forces us to rethink our assumptions about a foundational anthropological concept: context. The goal is to defamiliarize the conventional sign by revealing how it is just one of several semiotic modalities and then to explore the very different nonsymbolic properties of those other semiotic forms that are usually occluded by and collapsed into the symbolic in anthropological analysis. An anthropology beyond the human is in large part about learning to appreciate how the human is also the product of that which lies beyond human contexts.

Those concerned with nonhumans have often tried to overcome the familiar Cartesian divide between the symbolic realm of human meanings and the meaningless realm of objects either by mixing the two—terms such as *natures-cultures* or *material-semiotic* are indicative of this—or by reducing one of these poles to the other. By contrast, "The Open Whole" aims to show that the recognition of representational processes as something unique to, and in a sense even synonymous with, life allows us to situate distinctively human ways of being in the world as both emergent from and in continuity with a broader living semiotic realm.

If, as I argue, the symbolic is "open," to what exactly does it open? Opening the symbolic, through this exploration of signs beyond the symbolic, forces us to ponder what we might mean by the "real," given that the hitherto secure foundations for the real in anthropology—the "objective" and the contextually constructed—are destabilized by the strange and hidden logics of those signs that emerge, grow, and circulate in a world beyond the human.

Chapter 2, "The Living Thought," considers the implications of the claim, laid out in chapter 1, that all beings, including those that are nonhuman, are constitutively semiotic. All life is semiotic and all semiosis is alive. In important ways, then, life and thought are one and the same: life thinks; thoughts are alive.

This has implications for understanding who "we" are. Wherever there are "living thoughts" there is also a "self." "Self," at its most basic level, is a product of semiosis. It is the locus—however rudimentary and ephemeral—of a living dynamic by which signs come to represent the world around them to a "someone" who emerges as such as a result of this process. The world is thus "animate." "We" are not the only kind of we.

The world is also "enchanted." Thanks to this living semiotic dynamic, mean-ing (i.e., means-ends relations, significance, "aboutness," telos) is a constitutive feature of the world and not just something we humans impose on it. Appreciating life and thought in this manner changes our understanding of what selves are and how they emerge, dissolve, and also merge into new kinds of we as they interact with the other beings that make the tropical forest their home in that complex web of relations that I call an "ecology of selves."

The way Runa struggle to comprehend and enter this ecology of selves amplifies and makes apparent the peculiar logic of association by which living thoughts relate. If, as Strathern (1995) has argued, anthropology is at base about "the Relation," understanding some of the strange logics of association that emerge in this ecology of selves has important implications for our discipline. As we will see, it reveals how indistinction figures as a central aspect of relating. This changes our understandings of relationality; difference no longer sits so easily at the foundation of our conceptual framework, and this changes how we think about the central role that alterity plays in our discipline. A focus on this living semiotic dynamic in which indistinction (not to be confused with intrinsic similarity) operates also helps us see how "kinds" emerge in the world beyond the human. Kinds are not just human mental categories, be these innate or conventional; they result from how beings relate to each other in an ecology of selves in ways that involve a sort of confusion.

Just how to go about relating to those different beings that inhabit this vast ecology of selves poses pragmatic as well as existential challenges. Chapters 3 and 4 examine ethnographically how the Runa deal with such challenges, and these chapters reflect, more generally, on what we can learn from this.

Chapter 3, "Soul Blindness," is about the general problem of how death is intrinsic to life. Hunting, fishing, and trapping place the Runa in a particular relationship with the many beings that make up the ecology of selves in which they live. These activities force the Runa to assume their points of view, and indeed to recognize that all these creatures that they hunt, as well as the many other creatures with which those hunted animals relate, have points of view. It forces them to recognize that these creatures inhabit a network of relations that is predicated in part on the fact that its constitutive members are living, thinking selves. The Runa enter this ecology of selves as selves. They hold that their ability to enter this web of relations—to be aware of and to relate to other selves—depends on the fact that they share this quality with the other beings that make up this ecology.

Being aware of the selfhood of the many beings that people the cosmos poses particular challenges. The Runa enter the forest's ecology of selves in order to hunt, which means that they recognize others as selves like them- selves in order to turn them into nonselves. Objectification, then, is the flipside of animism, and it is not a straightforward process. Furthermore, one's ability to destroy other selves rests on and also highlights the fact that one is an ephemeral self—a self that can all too quickly cease being a self. Under the rubric "soul blindness," this chapter charts moments where this ability to rec- ognize other selves is lost and how this results in a sort of monadic alienation as one is, as a consequence, avulsed from the relational ecology of selves that constitutes the cosmos.

That death is intrinsic to life exemplifies something Cora Diamond (2008) calls a "difficulty of reality." It is a fundamental contradiction that can over- whelm us with its incomprehensibility. And this difficulty, as she emphasizes, is compounded by another one: such contradictions are at times, and for some, completely unremarkable. The feeling of disjunction that this creates is also part of the difficulty of reality. Hunting in this vast ecology of selves in which one must stand as a self in relation to so many other kinds of selves who one then tries to kill brings such difficulties to the fore; the entire cosmos reverberates with the contradictions intrinsic to life.

This chapter, then, is about the death in life, but it is especially about something Stanley Cavell calls the "little deaths" of "everyday life" (Cavell 2005: 128). There are many kinds and scales of death. There are many ways in which we cease being selves to ourselves and to each other. There are many ways of being pulled out of relation and many occasions where we turn a blind eye to and even kill relation. There are, in short, many modalities of disenchantment. At times the horror of this everyday fact of our existence bursts into our lives, and thus becomes a difficulty of reality. At others it is simply ignored.

Chapter 4, "Trans-Species Pidgins," is the second of these two chapters concerned with the challenges posed by living in relation to so many kinds of selves in this vast ecology of selves. It focuses on the problem of how to safely and successfully communicate with the many kinds of beings that people the cosmos. How to understand and be understood by beings whose grasp of human language is constantly in question is difficult in its own right. And when successful, communication with these beings can be destabilizing. Communication, to an extent, always involves communion. That is, communicating with others entails some measure of what Haraway (2008) calls "becoming with" these others. Although this promises to widen ways of being, it can also be very threatening to a more distinctly human sense of self that the Runa, despite this eagerness for expansion, also struggle to maintain. Accordingly, people in Ávila find creative strategies to open channels of communication with other beings in ways that also put brakes on these transgressive processes that can otherwise be so generative.

Much of this chapter focuses on the semiotic analysis of human attempts to understand and be understood by their dogs. For example, people in Ávila struggle to interpret their dogs' dreams, and they even give their dogs hallucinogens in order to be able to give them advice—in the process shifting to a sort of trans-species pidgin with unexpected properties.

The human-dog relation is special in part because of the way it links up to other relations. With and through their dogs people connect both to the broader forest ecology of selves and to an all-too-human social world that stretches beyond Ávila and its surrounding forests and that also catches up layers of colonial legacies. This chapter and the two that follow consider relationality in this expanded sense. They are concerned not just with how the Runa relate to the forest's living creatures but also with how the Runa relate to its spirits as well as to the many powerful human beings who have left their traces on the landscape.

How the Runa relate to their dogs, to the living creatures of the forest, to its ethereal but real spirits, and to the various other figures—the estate bosses, the priests, the colonists—that over the course of time have come to people their world cannot be distentangled. They are all part of this ecology that makes the Runa who they are. Nonetheless, I resist the temptation to treat this relational knot as an irreducible complexity. There is something we can learn about all these relations—and relationality more broadly—by paying careful attention to the specific modalities through which communication is attempted with different kinds of beings. These struggles to communicate reveal certain formal properties of relation—a certain logic of association, a set of constraints—that are neither the contingent products of earthly biologies nor those of human histories but which are instantiated in, and thus give shape to, both.

The property that most interests me here is hierarchy. The life of signs is characterized by a host of unidirectional and nested logical properties— properties that are consummately hierarchical. And yet, in the hopeful politics we seek to cultivate, we privilege heterarchy over hierarchy, the rhizomatic over the arborescent, and we celebrate the fact that such horizontal processes—lateral gene transfer, symbiosis, commensalism, and the like—can be found in the nonhuman living world. I believe this is the wrong way to ground politics. Morality, like the symbolic, emerges within—not beyond—the human. Projecting our morality, which rightfully privileges equality, on a relational landscape composed in part of nested and unidirectional associations of a logical and ontological, but not a moral, nature is a form of anthropocentric narcissism that renders us blind to some of the properties of that world beyond the human. As a consequence it makes us incapable of harnessing them politically. Part of the interest of this chapter, then, lies in charting how such nested relations get caught up and deployed in moral worlds without themselves being the products of those moral worlds.

The fifth chapter, "Form's Effortless Efficacy," is the place where I flesh out this account—to which I have heretofore been alluding—of the anthropological significance of form. That is, it is about how specific configurations of limits on possibility emerge in this world, the peculiar manner in which these redundancies propagate, and the ways in which they come to matter to lives, human and otherwise, in the forests around Ávila.

Form is difficult to treat anthropologically. Neither mind nor mechanism, it doesn't easily fit the dualistic metaphysics we inherit from the Enlightenment—a metaphysics that even today, in ways we may not necessarily always notice, steers us toward seeing cause in terms either of mechanistic pushes and pulls or of the meanings, purposes, and desires that we have generally come to relegate to the realm of the human. Much of the book so far has been concerned with dismantling some of the more persistent legacies of this dualism by tracing the implications of recognizing that meaning, broadly defined, is part and parcel of the living world beyond the human. This chapter, by contrast, seeks to further this endeavor by going beyond not only the human but also life. It is about the strange properties of pattern propagation that exceed life despite the fact that such patterns are harnessed, nurtured, and amplified by life. In a tropical forest teeming with so many forms of life these patterns proliferate to an unprecedented degree. To engage with the forest on its terms, to enter its relational logic, to think with its thoughts, one must become attuned to these.

By "form" here, I'm not, then, referring to the conceptual structures—innate or learned—through which we humans apprehend the world, nor am I referring to an ideal Platonic realm. Rather, I am referring to a strange but nonetheless worldly process of pattern production and propagation, a process Deacon (2006, 2012) characterizes as "morphodynamic"—one whose peculiar generative logic necessarily comes to permeate living beings (human and nonhuman) as they harness it.

Even though form is not mind it is not thinglike either. Another difficulty for anthropology is that form lacks the tangible otherness of a standard ethnographic object. When one is inside it there is nothing against which to push; it cannot be defined by the way it resists. It is not amenable to this kind of palpation, to this way of knowing. It is also fragile and ephemeral. Like the vortices of the whirlpools that sometimes form in the swift-flowing Amazonian headwaters, it simply vanishes when the special geometry of constraints that sustains it disappears. It thus remains largely hidden from our standard modes of analysis.

Through the examination of a variety of ethnographic, historical, and biological examples summoned together in an attempt to make sense of a puzzling dream I had about my relation to some of the animals of the forest and the spirit masters that control them, this chapter tries to understand some of the peculiar properties of form. It tries to understand the ways form does something to cause-and-effect temporality and the ways it comes to exhibit its own kind of "effortless efficacy" as it propagates itself through us. I am particularly interested here in how the logic of form affects the logic of living thoughts. What happens to thought when it is freed from its own intentions, when, in Lévi-Strauss's words, we ask of it no return (Lévi-Strauss 1966: 219)? What kinds of ecologies does it sound, and, in the process, what new kinds of relations does it make possible?

This chapter is also, nonetheless, concerned with the very practical problem of getting inside form and doing something with it. The wealth of the forest— be it game or extractive commodities—accumulates in a patterned way. Accessing it requires finding ways to enter the logic of these patterns. Accordingly, this chapter also charts the various techniques, shamanic and otherwise, used to do this, and it also attends to the painful sense of alienation the Runa feel when they are unable to enter the many new forms that have come over time to serve as the reservoirs for so much power and wealth.

Rethinking cause through form forces us to rethink agency as well. What is this strange way of getting something done without doing anything at all? What kinds of politics can come into being through this particular way of creating associations? Grasping how form emerges and propagates in the forest and in the lives of those who relate to it—be they river dolphins, hunters, or rubber bosses—and understanding something about form's effortless efficacy is central to developing an anthropology that can attend to those many processes central to life, human and nonhuman, which are not built from quanta of difference.

How Forests Think is a book, ultimately, about thought. It is, to quote Viveiros de Castro, a call to make anthropology a practice for "la décolonisation permanente de la pensée" (Viveiros de Castro 2009: 4). My argument is that we are colonized by certain ways of thinking about relationality. We can only imagine the ways in which selves and thoughts might form associations through our assumptions about the forms of associations that structure human language. And then, in ways that often go unnoticed, we project these assumptions onto nonhumans. Without realizing it we attribute to nonhumans properties that are our own, and then, to compound this, we narcissistically ask them to provide us with corrective reflections of ourselves.

So, how should we think with forests? How should we allow the thoughts in and of the nonhuman world to liberate our thinking? Forests are good to think because they themselves think. Forests think. I want to take this seriously, and

I want to ask, What are the implications of this claim for our understandings of what it means to be human in a world that extends beyond us?

Wait. How can I even make this claim that forests think? Shouldn't we only ask how people think forests think? I'm not doing this. Here, instead, is my provocation. I want to show that the fact that we can make the claim that forests think is in a strange way a product of the fact that forests think. These two things—the claim itself and the claim that we can make the claim—are related: It is because thought extends beyond the human that we can think beyond the human.

This book, then, aims to free our thinking of that excess conceptual baggage that has accumulated as a result of our exclusive attention—to the neglect of everything else—to that which makes us humans exceptional. *How Forests Think* develops a method for crafting new conceptual tools out of the unexpected properties of the world beyond the human that we discover ethnographically. And in so doing it seeks to liberate us from our own mental enclosures. As we learn to attend ethnographically to that which lies beyond the human, certain strange phenomena suddenly come to the fore, and these strange phenomena amplify, and in the process come to exemplify, some of the general properties of the world in which we live. If through this form of analysis we can find ways to further amplify these phenomena, we can then cultivate them as concepts and mobilize them as tools. By methodologically privileging amplification over, say, comparison or reduction we can create a somewhat different anthropology, one that can help us understand how we might better live in a world we share with other kinds of lives.

The logics of living dynamics, and the sorts of ancillary phenomena these both create and catch up, might at first appear strange and counterintuitive. But, as I hope to show, they also permeate our everyday lives, and they might help us understand our lives differently if we could just learn to listen for them. This emphasis on defamiliarization—coming to see the strange as familiar so that the familiar appears strange—calls to mind a long anthropological tradition that focuses on how an appreciation for context (historical, social, cultural) destabilizes what we take to be natural and immutable modes of being. And yet, when compared to the distance-making practices associated with more traditional liberatory ethnographic or genealogical exercises, seeing the human from somewhat beyond the human does not merely destabilize the taken for granted; it changes the very terms of analysis and comparison.

This reach beyond the human changes our understanding of foundational analytical concepts such as context but also others, such as representation, relation, self, ends, difference, similarity, life, the real, mind, person, thought, form, finitude, future, history, cause, agency, relation, hierarchy, and generality. It changes what we mean by these terms and where we locate the phenomena to which they refer, as well as our understanding of the effects such phenomena have in the living world in which we live.

The final chapter, "The Living Future (and the Imponderable Weight of the Dead)," builds on this way of thinking with forests that I develop in this book as it takes as its focus another enigmatic dream, in this case one of a hunter who is not sure if he is the rapacious predator (who appears here as a white policeman) or the helpless prey of his oneiric prophecy. The interpretive dilemma that this dream poses, and the existential and psychic conflict that it thus lays bare, concerns how to continue as a self and what such continuity might mean in the ecology of selves in which the Runa live—an ecology that is firmly rooted in a forest realm that reaches well beyond the human but which also catches up in its tendrils the detritus of so many all-too-human pasts. This chapter, more broadly, is about survival. That is, it is about the relation of continuity and growth to absence. Ethnographic attention to the problem of survival in the particular colonially inflected ecology of selves in which the Runa live tells us something more general about how we might become new kinds of *we*, in relation to such absences, and how, in this process, "we" might, to use Haraway's (2008) term, "flourish."

Understanding this dream and what it can tell us about survival calls for a shift, not only regarding anthropology's object—the human—but also regarding its temporal focus. It asks us to recognize more generally how life—human and nonhuman—is not just the product of the weight of the past on the present but how it is also the product of the curious and convoluted ways in which the future comes to bear upon a present.

That is, all semiotic processes are organized around the fact that signs represent a future possible state of affairs. The future matters to living thoughts. It is a constitutive feature of any kind of self. The life of signs is not, then, just in the present but also in a vague and possible future. Signs are oriented toward the ways in which future signs will likely represent their relationship to a likely state of affairs. Selves, then, are characterized by what Peirce calls a "being *in futuro*" (CP 2.86), or a "living future" (CP 8.194).[11] This particular kind of causality, whereby a future comes to affect the present via the mediation of signs, is unique to life.

In the life of signs future is also closely related to absence. All kinds of signs in some way or other re-present what is not present. And every successful representation has another absence at its foundation; it is the product of the history of all the other sign processes that less accurately represented what would be. What one is as a semiotic self, then, is constitutively related to what one is not. One's future emerges from and in relation to a specific geometry of absent histories. Living futures are always "indebted" to the dead that surround them.

At some level this way in which life creates future in negative but constitutive relation to all its pasts is characteristic of all semiotic processes. But it is a dynamic that is amplified in the tropical forest, with its unprecedented layers of mutually constitutive representational relationships. Runa engagements with this complex ecology of selves create even more future.

Chapter 6, then, is primarily concerned with one particular manifestation of this future: the realm of the afterlife located deep in the forest and inhabited by the dead and the spirit masters that control the forest's animals. This realm is the product of the relationship that invisible futures have to the painful histories of the dead that make life possible. Around Ávila these dead take the form of were-jaguars, masters, demons, and the specters of so many pre-Hispanic, colonial, and republican pasts; all these continue, in their own ways, to haunt the living forest.

This chapter traces how this ethereal future realm relates to the concrete one of everyday Runa existence. The Runa, living in relation to the forest's vast ecology of selves, also live their lives with one foot *in futuro*. That is, they live their lives with one foot in the spirit realm that is the emergent product of the ways in which they engage with the futures and the pasts that the forest comes to harbor in its relational webs. This other kind of "beyond," this *after*-life, this *super*-nature, is not exactly natural (or cultural), but it is nonetheless real. It is its own kind of irreducible real, with its own distinctive properties and its own tangible effects in a future present.

The fractured and yet necessary relationship between the mundane present and the vague future plays out in specific and painful ways in what Lisa Stevenson (2012; see also Butler 1997) might call the psychic life of the Runa self, immersed and informed as it is by the ecology of selves in which it lives. The Runa are both of and alienated from the spirit world, and survival requires cultivating ways to allow something of one's future self—living tenuously in the spirit realm of the forest masters—to look back on and call out to that more mundane part of oneself that might then hopefully respond. This ethereal realm of continuity and possibility is the emergent product of a whole host of trans-species and transhistorical relations. It is the product of the imponderable weight of the many dead that make a living future possible.

That hunter's challenge of surviving as an *I*, as it was revealed in his dream and as it plays out in this ecology of selves, depends on how he is hailed by others—others that may be human or nonhuman, fleshly or virtual. It also depends on how he responds. Is he the white policeman who might turn on his Runa neighbors with a blood thirst that terrifies him? Is he helpless prey? Or might he not be a runa puma, a were-jaguar, capable, even, of returning a jaguar's gaze?

Let this runa puma, this one who both is and is not us, be, like Dante's Virgil, our guide as we wander this "dense and difficult" forest—this "selva selvaggia" where words so often fail us. Let this runa puma guide us with the hope that we too may learn another way to attend and respond to the many lives of those selves that people this sylvatic realm.

ONE

The Open Whole

By a feeling I mean an instance of that sort of element of consciousness which is all that it is positively, in itself, regardless of anything else. . . . [A] feeling is absolutely simple and without parts—as it evidently is, since it is whatever it is regardless of anything else, and therefore regardless of any part, which would be something other than the whole.

—Charles Peirce, *The Collected Papers* 1.306–10

One evening while the grown-ups gathered around the hearth drinking manioc beer, Maxi, settling back to a quieter corner of the house, began to tell his teenage neighbor Luis and me about some of his recent adventures and mishaps. Fifteen or so and just beginning to hunt on his own, he told us of the day he stood out in the forest for what seemed an eternity, waiting for something to happen, and how, all of a sudden, he found himself close to a herd of collared peccaries moving through the underbrush. Frightened, he hoisted himself into the safety of a little tree and from there fired on and hit one of the pigs. The wounded animal ran off toward a little river and . . . "*tsupu*."

Tsupu. I've deliberately left Maxi's utterance untranslated. What might it mean? What does it sound like?

Tsupu, or *tsupuuuʰ*, as it is sometimes pronounced, with the final vowel dragged out and aspirated, refers to an entity as it makes contact with and then penetrates a body of water; think of a big stone heaved into a pond or the compact mass of a wounded peccary plunging into a river's pool. *Tsupu* probably did not immediately conjure such an image (unless you speak lowland Ecuadorian Quichua). But what did you feel upon learning what it describes? Once I tell people what *tsupu* means, they often experience a sudden feel for its meaning: "Oh, of course, *tsupu!*"

By contrast, I would venture that even after learning that the greeting *"causanguichu,"* used when encountering someone who hasn't been seen in a long time, means "Are you still alive?" you don't have such a feeling. *Causanguichu* certainly feels like what it means to native speakers of Quichua, and over the years I too have come to develop a feel for its meaning. But what is it about *tsupu* that causes its meaning to feel so evident even for many people who don't speak Quichua? *Tsupu* somehow feels like a pig plunging into water.

How is it that *tsupu* means? We know that a word like *causanguichu* means by virtue of the ways in which it is inextricably embedded, through a dense historically contingent tangle of grammatical and syntactic relations, with other such words in that uniquely human system of communication we call language. And we know that what it means also depends on the ways in which language is itself caught up in broader social, cultural, and political contexts, which share similar historically contingent systemic properties. In order to develop a feel for *causanguichu* we have to grasp something of the totality of the interrelated network of words in which it exists. We also need to grasp something of the broader social context in which it is and has been used. Making sense of how we live inside these kinds of changing contexts that we both make and that make us has long been an important goal of anthropology. For anthropology the "human," as a being and an object of knowledge, emerges only by attending to how we are embedded in these uniquely human contexts—these "complex wholes" as E.B. Tylor's (1871) classic definition of culture terms them.

But if *causanguichu* is firmly in language, *tsupu* seems somehow outside it. *Tsupu* is a sort of paralinguistic parasite on the language that somewhat indifferently bears it. *Tsupu* is, in a way, as Peirce might say, "all that it is positively, in itself, regardless of anything else." And this admittedly minor fact, that this strange little quasi-word is not quite made by its linguistic context, troubles the anthropological project of making sense of the human via context.

Take *causanguichu*'s root, the lexeme *causa-*, which is marked for person and inflected by a suffix that signals its status as a question:

causa-ngui-chu
live-2-INTER[1]
Are you still alive?

Through its grammatical inflections *causanguichu* is inextricably related to the other words that make up the Quichua language. *Tsupu*, by contrast, doesn't really interact with other words and therefore can't be modified to reflect any such possible relations. Being "all that it is positively in itself," it can't even be grammatically negated. What kind of thing, then, is *tsupu*? Is it even a word? What does its anomalous place in language reveal about language? And what can it tell us about the anthropological project of grasping the various ways in which linguistic as well as sociocultural and historical contexts form the conditions of possibility both for human life and for our ways of attending to it?

Although not exactly a word, *tsupu* certainly is a sign. That is, it certainly is, as the philosopher Charles Peirce put it, "something which stands to somebody for something in some respect or capacity" (CP 2.228). This is quite different from Saussure's (1959) more humanist treatment of signs with which we anthropologists tend to be more familiar. For Saussure human language is the paragon and model for all sign systems (1959: 68). Peirce's definition of a sign, by contrast, is much more agnostic about what signs are and what kinds of beings use them; for him not all signs have languagelike properties, and, as I discuss below, not all the beings who use them are human. This broader definition of the sign helps us become attuned to the life signs have beyond the human as we know it.

Tsupu captures to some extent and in some particular way something of a pig plunging into water, and it does so—weirdly—not just for Quichua speakers, but to some degree for those of us who may not have any familiarity with the language that carries it along.[2] What might paying attention to this not-quite-wordlike-kind-of-sign reveal? Feeling *tsupu*, "in itself, regardless of anything else," can tell us something important about the nature of language and its unexpected openings toward the world "itself." And insofar as it can help us understand how signs are not just bounded by human contexts, but how they also reach beyond them. Insofar, that is, as it can help reveal how signs are also in, of, and about other sensuous worlds that we too can feel, it can also tell us something about how we can move beyond understanding the human in terms of the "complex wholes" that make us who we are. In sum, appreciating what it might mean "to live" (Quichua *causa-ngapa*) in worlds that are open to that which extends beyond the human might just allow us to become a little more "worldly."[3]

IN AND OF THE WORLD

In uttering "tsupu," Maxi brought home something that happened in the forest. Insofar as Luis, or I, or you, feel tsupu we come to grasp something of Maxi's experience of being near a wounded pig plunging into a pool of water. And we can come to have this feeling even if we weren't in the forest that day. All signs, and not just tsupu, are in some way or another about the world in this sense. They "re-present." They are about something not immediately present.

But they are also all, in some way or another, in and of the world. When we think of situations in which we use signs to represent an event, such as the one I've just described, this quality may be hard to see. Sitting back in a dark corner of a thatched roof house listening to Maxi talk about the forest is not the same as having been present to that pig plunging into water. Isn't this "radical discontinuity" with the world another important hallmark of signs?[4] Insofar as signs do not provide any sort of immediate, absolute, or certain purchase on the entities they represent, it certainly is. But the fact that signs always mediate does not mean that they also necessarily exist in some separate domain inside (human) minds and cut off from the entities they stand for. As I will show, they are not just about the world. They are also in important ways in it.

Consider the following. Toward the end of a day spent walking in the forest, Hilario, his son Lucio, and I came upon a troop of woolly monkeys moving through the canopy. Lucio shot and killed one, and the rest of the troop dispersed. One young monkey, however, became separated from the troop. Finding herself alone she hid in the branches of an enormous red-trunked tree that poked out of the forest canopy high above.[5]

In the hope of startling the monkey into moving to a more visible perch so that his son could shoot it Hilario decided to fell a nearby palm tree:

look out!
ta ta
I'll make it go *pu oh*
watch out![6]

Ta ta and *pu oh*, like *tsupu*, are images that sound like what they mean. *Ta ta* is an image of chopping: tap tap. *Pu oh* captures the process by which a tree falls. The snap that initiates its toppling, the swish of the crown free-falling through layers of forest canopy, and the crash and its echoes as it hits the ground are all enfolded in this sonic image.

Hilario then went and did what he said. He walked off a little way and with his machete began chopping rhythmically at a palm tree. The tapping of steel against trunk is clearly audible on the recording I made in the forest that afternoon (*ta ta ta ta . . .*)—as was the palm crashing down (*pu oh*).

Lowland Quichua has hundreds of "words" like *ta ta, pu oh,* and *tsupu* that mean by virtue of the ways in which they sonically convey an image of how an action unfolds in the world. They are ubiquitous in speech, especially in forest talk. A testament to their importance to Runa ways of being in the world is that the linguistic anthropologist Janis Nuckolls (1996) has written an entire book—titled, appropriately, *Sounds Like Life*—about them.

A "word" such as *tsupu* is like the entity it represents thanks to the ways in which the differences between the "sign vehicle" (i.e., the entity that is taken as a sign, in this case the sonic quality of *tsupu*)[7] and the object (in this case the plunging-into-water that this "word" simulates) are ignored.[8] Peirce called these kinds of signs of likeness "icons." They conform to the first of his three broad classes of signs.

As Hilario had anticipated, the sound of the palm tree crashing frightened the monkey from her perch. This event itself, and not just its before-the-fact imitation, can also be taken as a kind of sign. It is a sign in the sense that it too came to be "something which stands to somebody for something in some respect or capacity." In this case the "somebody" to whom this sign stands is not human. The palm crashing down stands for something to the monkey. Significance is not the exclusive province of humans because we are not the only ones who interpret signs. That other kinds of beings use signs is one example of the ways in which representation exists in the world beyond human minds and human systems of meaning.

The palm crashing down becomes significant in a way that differs from its imitation *pu oh*.[9] *Pu oh* is iconic in the sense that it, in itself, is in some respect like its object. That is, it functions as an image when we fail to notice the differences between it and the event that it represents. It means due to a certain kind of absence of attention to difference. By ignoring the myriad characteristics that make any entity unique, a very restricted set of characteristics is amplified, here by virtue of the fact that the sound that simulates the action also happens to share these characteristics.

The crashing palm itself comes to signify something for the monkey in another capacity. The crash, as sign, is not a likeness of the object it represents. Instead, it points to something else. Peirce calls this sort of sign an "index." Indices constitute his second broad class of signs.

Before exploring indices further, I want to briefly introduce the "symbol"—Peirce's third kind of sign. Unlike iconic and indexical modes of reference, which form the bases for all representation in the living world, symbolic reference is, on this planet at least, a form of representation that is unique to humans. Accordingly, as anthropologists of the human we are most familiar with its distinctive properties. Symbols refer, not simply through the similarity of icons, or solely through the pointing of indices. Rather, as with the word *causanguichu*, they refer to their object indirectly by virtue of the ways in which they relate systemically to other such symbols. Symbols involve convention. This is why *causanguichu* only means—and comes to feel meaningful—by virtue of the established system of relationships it has with other words in Quichua.

The palm that Hilario sent crashing down that afternoon startled the monkey. As an index it forced her to notice that something just happened, even though what just happened remained unclear.[10] Whereas icons involve not noticing, indices focus the attention. If icons are what they are "in themselves" regardless of the existence of the entity they represent, indices involve facts "themselves." Whether or not someone was there to hear it, whether or not the monkey, or anyone else for that matter, took this occurrence to be significant, the palm, itself, still came crashing down.

Unlike icons, which represent by virtue of the resemblances they share with objects, indices represent "by virtue of real connections to them" (Peirce 1998c: 461; see also CP 2.248). Tugging on the stems of woody vines, or lianas, that extend up into the canopy is another strategy to scare monkeys out of their hidden perches (see frontispiece, this chapter). To the extent that such an action can startle a monkey it is because of a chain of "real connections" among disparate things: the hunter's tug is transmitted, via the liana, high up to the tangled mat of epiphytes, lianas, moss, and detritus that accumulates to form the perch atop which the hiding monkey sits.

Although one might say that the hunter's tug, propagated through the liana and mat, literally shakes the monkey out of her sense of security, how this monkey comes to take this tug as a sign cannot be reduced to a deterministic chain of causes and effects. The monkey need not necessarily perceive the shaking perch to be a sign of anything. And in the event that she does, her reaction will be something other than the effect of the force of the tug propagated up the length of the liana.

Indices involve something more than mechanical efficiency. That something more is, paradoxically, something less. It is an absence. That is, to the extent that indices are noticed they impel their interpreters to make connections between some event and another potential one that has not yet occurred.

A monkey takes the moving perch, as sign, to be connected to something else, for which it stands. It is connected to something dangerously different from her present sense of security. Maybe the branch she is perched on is going to break off. Maybe a jaguar is climbing up the tree ... Something is about to happen, and she had better do something about it. Indices provide information about such absent futures. They encourage us to make a connection between what is happening and what might potentially happen.

LIVING SIGNS

Asking whether signs involve sound images like *tsupu*, or whether they come to mean through events like a palm crashing down, or whether their sense emerges in some more systemic and distributed manner, like the interrelated network of words printed on the pages that make up this book, might encourage us to think about signs in terms of the differences in their tangible qualities. But signs are more than things. They don't squarely reside in sounds, events, or words. Nor are they exactly in bodies or even minds. They can't be precisely located in this way because they are ongoing relational processes. Their sensuous qualities are only one part of the dynamic through which they come to be, to grow, and to have effects in the world.

In other words signs are alive. A crashing palm tree—taken as sign—is alive insofar as it can grow. It is alive insofar as it will come to be interpreted by a subsequent sign in a semiotic chain that extends into the possible future.

The startled monkey's jump to a higher perch is a part of this living semiotic chain. It is what Peirce called an "interpretant," a new sign that interprets the way in which a prior sign relates to its object.[11] Interpretants can be further specified through an ongoing process of sign production and interpretation that increasingly captures something about the world and increasingly orients an interpreting self toward this aboutness. Semiosis is the name for this living sign process through which one thought gives rise to another, which in turn gives rise to another, and so on, into the potential future.[12] It captures the way in which living signs are not just in the here and now but also in the realm of the possible.

Although semiosis is something more than mechanical efficiency, thinking is not just confined to some separate realm of ideas.[13] A sign has an effect, and this, precisely, is what an interpretant is. It is the "proper significate effect that the sign produces" (CP 5.475). The monkey's jump, sparked by her reaction to a crashing palm, amounts to an interpretant of a prior sign of danger. It makes visible an energetic component that is characteristic of all sign processes, even those that might seem purely "mental."[14] Although semiosis is something more than energetics and materiality, all sign processes eventually "do things" in the world, and this is an important part of what makes them alive.[15]

Signs don't come from the mind. Rather, it is the other way around. What we call mind, or self, is a product of semiosis. That "somebody," human or non-human, who takes the crashing palm to be significant is a "self that is just coming into life in the flow of time" (CP 5.421) by virtue of the ways in which she comes to be a locus—however ephemeral—for the "interpretance" of this sign and many others like it. In fact, Peirce coined the cumbersome term *interpretant* to avoid the "homunculus fallacy" (see Deacon 2012: 48) of seeing a self as a sort of black box (a little person inside us, a homunculus) who would be the interpreter of those signs but not herself the product of those signs. Selves, human or nonhuman, simple or complex, are outcomes of semiosis as well as the starting points for new sign interpretation whose outcome will be a future self. They are waypoints in a semiotic process.

These selves, "just coming into life," are not shut off from the world; the semiosis occurring "inside" the mind is not intrinsically different from that which occurs among minds. That palm crashing down in the forest illustrates this living worldly semiosis as it is embedded in an ecology of disparate emerging selves. Hilario's iconic simulation of a falling palm charts a possible future that then becomes realized in a palm that he actually fells. Its crash, in turn, is interpreted by another being whose life will change thanks to the way she takes this as a sign of something upon which she must act. What emerges is a highly mediated but nevertheless unbroken chain that jumps from the realm of human speech to that of human bodies and their actions, and from these to events-in-the-world such as a tree crashing down that these realized embodied intentions actualize, and from here to the equally physical reaction that the semiotic interpretation of this event provokes in another kind of primate high up in a tree. The crashing palm and the human who felled it came to affect the monkey, notwithstanding their physical separation from her. Signs have worldly effects even though they are not reducible to physical cause-and-effect.

Such tropical trans-species attempts at communication reveal the living worldly nature of semiosis. All semiosis (and by extension thought) takes place in minds-in-the-world. To highlight this characteristic of semiosis this is how

Peirce described the thought practices of Antoine Lavoisier, the eighteenth-century French aristocrat and founder of the modern field of chemistry:

> Lavoisier's method was . . . to dream that some long and complicated chemical pro-cess would have a certain effect, to put it into practice with dull patience, after its inevitable failure, to dream that with some modification it would have another result, and to end by publishing the last dream as a fact: his way was to carry his mind into his laboratory, and literally to make of his alembics and cucurbits instru-ments of thought, giving a new conception of reasoning as something which was to be done with one's eyes open, in manipulating real things instead of words and fancies. (CP 5.363)

Where would we locate Lavoisier's thoughts and dreams? Where, if not in this emerging world of blown glass cucurbits and alembics and the mixtures contained in their carefully delimited spaces of absence and possibility, is his mind, and future self, coming in to being?

ABSENCES

Lavoisier's blown glass flasks point to another important element of semiosis. Like these curiously shaped receptacles, signs surely have an important mate-riality: they possess sensuous qualities; they are instantiated with respect to the bodies that produce and are produced by them; and they can make a dif-ference in the worlds that they are about. And yet, like the space delimited by the walls of the flask, signs are also in important ways immaterial. A glass flask is as much about what it is as it is about what it is not; it is as much about the vessel blown into form by the glassmaker—and all the material qualities and technological, political, and socioeconomic histories that made that act of creation possible—as it is about the specific geometry of absence that it comes to delimit. Certain kinds of reactions can take place in that flask because of all the others that are excluded from it.

This kind of absence is central to the semiosis that sustains and instantiates life and mind. It is apparent in what played out in the forest that afternoon as we were out hunting monkeys. Now that that young woolly monkey had moved to a more exposed perch Lucio tried to shoot at it with his muzzle-loading black powder shotgun. But when he pulled the trigger the hammer simply clicked down on the firing cap. Lucio quickly replaced the defective cap and reloaded—this time packing the barrel with an extra dose of lead shot. When the monkey climbed to an even more exposed position, Hilario encouraged his

FIGURE 4. A muzzle-loading shotgun *(illapa)*. Photo by author.

son to fire again: "Hurry, now really!" Wary of the precarious nature of his firearm, however, Lucio first uttered, *"teeeye."*

Teeeye, like *tsupu, ta ta*, and *pu oh*, is an image in sound. It is iconic of a gun successfully firing and hitting its target. The mouth that pronounces it is like a flask that assumes the various shapes of a firing gun. First the tongue taps on the palette to produce the stopped consonant the way a hammer strikes a firing cap. Then the mouth opens ever wider as it pronounces the expanding elongated vowel, the way lead shot, propelled by the explosion of powder ignited by the cap, sprays out of the barrel (figure 4).

Moments later Lucio pulled the trigger. And this time, with a deafening *teeeye*, the gun fired.

Teeeye is, at many levels, a product of what it is not. The shape of the mouth effectively eliminates all the many other sounds that could have been made as breath is voiced. What is left is a sound that "fits" the object it represents thanks to the many sounds that are absent. The object that is not physically present constitutes a second absence. Finally, *teeeye* involves another absence in the sense that it is a representation of a future brought into the present in the hopes that this not-yet will affect the present. Lucio hopes his gun will successfully fire *teeeye* when he pulls the trigger. He imported this simulation into the present from the possible world that he hopes will come to be. This future-possible, which orients Lucio toward taking all the steps needed to make this future possible, is also a constitutive absence. What *teeeye* is—its significate effect, in short, its meaning—is dependent on all these things that it is not.

All signs, and not just those we might call magical, traffic in the future in the way that *teeeye* does. They are calls to act in the present through an absent but re-presented future that, by virtue of this call, can then come to affect the present; "Hurry, now really," as Hilario implored his son moments before he fired his gun, involves a prediction that there will still be an "it" up there to shoot. It is a call from the future as re-presented in the present.

Drawing inspiration from the ancient Chinese philosopher Lao-tzu and his reflection on how the hole at the hub is what makes a wheel useful, Terrence Deacon (2006) refers to the special kind of nothingness delimited by the spokes of a wheel, or by the glass of a flask, or by the shape of the mouth when uttering "*teeeye*" as a "constitutive absence." Constitutive absence, according to Deacon, is not just found in the world of artifacts or humans. It is a kind of relation to that which is spatially or temporally not present that is crucial to biology and to any kind of self (see Deacon 2012: 3). It points to the peculiar way in which, "in the world of mind, nothing—that which is *not*—can be a cause" (Bateson 2000a: 458, quoted in Deacon 2006).

As I discuss later in this chapter, and in subsequent ones as well, constitutive absence is central to evolutionary processes. That, for example, a lineage of organisms comes to increasingly fit a particular environment is the result of the "absence" of all the other lineages that were selected out. And all manner of sign processes, not just those associated directly with biological life, come to mean by virtue of an absence: iconicity is the product of what is not noticed; indexicality involves a prediction of what is not yet present; and symbolic reference, through a convoluted process that also involves iconicity and indexicality, points to and images absent worlds by virtue of the ways in which it is embedded in a symbolic system that constitutes the absent context for the meaning of any given word's utterance. In the "world of mind," constitutive absence is a particular mediated way in which an absent future comes to affect the present. This is why it is appropriate to consider telos—that future for the sake of which something in the present exists—as a real causal modality wherever there is life (see Deacon 2012).

The constant play between presence and these different kinds of absences gives signs their life. It makes them more than the effect of that which came before them. It makes them images and intimations of something potentially possible.

PROVINCIALIZING LANGUAGE

Considering crashing palms, jumping monkeys, and "words" like *tsupu* helps us see that representation is something both more general and more widely distributed than human language. It also helps us see that these other modes of representation have properties that are quite different from those exhibited by the symbolic modalities on which language depends. In short, considering those kinds of signs that emerge and circulate beyond the symbolic helps us see that we need to "provincialize" language.

My call to provincialize language alludes to Dipesh Chakrabarty's *Provincializing Europe* (2000), his critical account of how South Asian and South Asianist scholars rely on Western social theory to analyze South Asian social realities. To provincialize Europe is to recognize that such theory (with its assumptions about progress, time, etc.) is situated in the particular European context of its production. Social theorists of South Asia, Chakrabarty argues, turn a blind eye to this situated context and apply such theory as if it were universal. Chakrabarty asks us to consider what kind of theory might emerge from South Asia, or from other regions for that matter, once we circumscribe the European theory we once took as universal.

In showing that the production of a particular body of social theory is situated in a particular context and that there are other contexts for which this theory does not apply, Chakrabarty is making an implicit argument about the symbolic properties of the realities such theory seeks to understand. Context is an effect of the symbolic. That is, without the symbolic we would not have linguistic, social, cultural, or historical contexts as we understand them. And yet this kind of context does not fully create or circumscribe our realities because we also live in a world that exceeds the symbolic, and this is something our social theory must also find ways to address.

Chakrabarty's argument, then, is ultimately couched within humanist assumptions about social reality and the theory one might develop to attend to it, and so, if taken literally, its application to an anthropology beyond the human is limited. Nonetheless, I find provincialization useful metaphorically as a reminder that symbolic domains, properties, and analytics are always circumscribed by and nested within a broader semiotic field.

We need to provincialize language because we conflate representation with language and this conflation finds its way into our theory. We universalize this distinctive human propensity by first assuming that all representation is something human and then by supposing that all representation has languagelike properties. That which ought to be delimited as something unique becomes instead the bedrock for our assumptions about representation.

We anthropologists tend to view representation as a strictly human affair. And we tend to focus only on symbolic representation—that uniquely human semiotic modality.[16] Symbolic representation, manifested most clearly in language, is conventional, "arbitrary," and embedded in a system of other such symbols, which, in turn, is sustained in social, cultural, and political contexts that have similar systemic and conventional properties. As I mentioned earlier, the representational system associated with Saussure, which is the implicit one that underlies so much of contemporary social theory, concerns itself only with this kind of arbitrary, conventional sign.

There is another reason why we need to provincialize language: we conflate language with representation even when we don't explicitly draw on language or the symbolic for our theoretical tools. This conflation is most evident in our assumptions about ethnographic context. Just as we know that words only acquire meanings in terms of the greater context of other such words to which they systemically relate, it is an anthropological axiom that social facts can't be understood except by virtue of their place in a context made up of other such facts. And the same applies for the webs of cultural meanings or for the network of contingent discursive truths as revealed by a Foucauldian genealogy.

Context understood in this way, however, is a property of human conventional symbolic reference, which creates the linguistic cultural and social realities that make us distinctively human. It doesn't fully apply in domains such as human-animal relations that are not completely circumscribed by the symbolic but are nevertheless semiotic. The kinds of representational modalities shared by all forms of life—modalities that are iconic and indexical—are not context-dependent the way symbolic modalities are. That is, such representational modalities do not function by means of a contingent system of sign relations—a context—the way symbolic modalities do. So in certain semiotic domains context doesn't apply, and even in those domains such as human ones where it does, such contexts, as we can see by attending to that which lies beyond the human, are, as I will show, permeable. In short, complex wholes are also open wholes—hence this chapter's title. And open wholes reach beyond the human—hence this anthropology beyond the human.

This conflation of representation with language—the assumption that all representational phenomena have symbolic properties—holds even for those kinds of projects that are explicitly critical of cultural, symbolic, or linguistic approaches. It is apparent in classical materialist critiques of the symbolic and the cultural. It is also apparent in more contemporary phenomenological approaches that turn to the bodily experiences we also share with nonhuman beings as a way to avoid anthropocentric mind talk (see Ingold 2000; Csordas 1999; Stoller 1997). It is also, I should note, apparent in Eduardo Viveiros de Castro's multinaturalism (discussed in detail in chapter 2). When Viveiros de Castro writes that "a perspective is not a representation because representations are a property of the mind or spirit, whereas the point of view is located in the body" (1998: 478), he is assuming that attention to bodies (and their natures) can allow us to side step the thorny issues raised by representation.

The alignment between humans, culture, the mind, and representation, on the one hand, and nonhumans, nature, bodies, and matter, on the other, remains stable even in posthuman approaches that seek to dissolve the boundaries that have been erected to construe humans as separate from the rest of the world. This is true of Deleuzian approaches, as exemplified, for example, by Jane Bennett (2010), that deny the analytical purchase of representation and telos altogether—since these are seen, at best, as exclusively human mental affairs.

This alignment is also evident in attempts in science and technology studies (STS), especially those associated with Bruno Latour, to equalize the imbalance between unfeeling matter and desiring humans by depriving humans of a bit of their intentionality and symbolic omnipotence at the same time that they confer on things a bit more agency. In his image of "speech impediments," for example, Latour attempts to find an idiom that might bridge the analytical gap between speaking scientists and their supposedly silent objects of study. "Better to have marbles in one's mouth, when speaking about scientists," he writes, "then to slip absent-mindedly from mute things to the indisputable word of the expert" (2004: 67). Because Latour conflates representation and human language his only hope to get humans and nonhumans in the same frame is to literally mix language and things—to speak with marbles in his mouth. But this solution perpetuates Cartesian dualism because the atomic elements remain either human mind or unfeeling matter, despite the fact that these are more thoroughly mixed than Descartes would have ever dreamed, and even if one claims that their mixture precedes their realization. This analytic of mixture creates little homunculi at all levels. The hyphen in Latour's (1993: 106) "natures-cultures" is the new pineal gland in the little Cartesian heads that this analytic unwittingly engenders at all scales. An anthropology beyond the human seeks to find ways to move beyond this analytic of mixture.

Erasing the divide between the human mind and the rest of the world, or, alternatively, striving for some symmetrical mixing between mind and matter, only encourages this gap to emerge again elsewhere. An important claim I make in this chapter, and an important foundation for the arguments to be developed in this book, is that the most productive way to overcome this dualism is not to do away with representation (and by extension telos, intentionality, "aboutness," and selfhood), or simply project human kinds of representation elsewhere, but to radically rethink what it is that we take representation to be. To do this, we need first to provincialize language. We need, in Viveiros de Castro's words, to "decolonize thought," in order to see that thinking is not necessarily circumscribed by language, the symbolic, or the human.

This involves reconsidering who in this world represents, as well as what it is that counts as representation. It also involves understanding how different kinds of representation work and how these different kinds of representation variously interact with each other. What sort of life does semiosis take beyond the trappings of internal human minds, beyond specifically human propensities, such as the ability to use language, and beyond those specifically human concerns that those propensities engender? An anthropology beyond the human encourages us to explore what signs look like beyond the human.

Is such an exploration possible? Or do the all-too-human contexts in which we live bar us from such an endeavor? Are we forever trapped inside our linguistically and culturally mediated ways of thinking? My answer is no: a more complete understanding of representation, which can account for the ways in which that exceptionally human kind of semiosis grows out of and is constantly in interplay with other kinds of more widely distributed representational modalities, can show us a more productive and analytically robust way out of this persistent dualism.

We humans are not the only ones who do things for the sake of a future by re-presenting it in the present. All living selves do this in some way or another. Representation, purpose, and future are in the world—and not just in that part of the world that we delimit as human mind. This is why it is appropriate to say that there is agency in the living world that extends beyond the human. And yet reducing agency to cause and effect—to "affect"—side steps the fact that it is human and nonhuman ways of "thinking" that confer agency. Reducing agency to some sort of generic propensity shared by humans and nonhumans (which in such approaches includes objects) thanks to the fact that these entities can all equally be represented (or that they can confound these representations), and that they then participate by virtue of this in some sort of very humanlike narrative, trivializes this thinking by failing to distinguish among ways of thinking and by indiscriminately applying distinctively human ways of thinking (based on symbolic representation) to any entity.

The challenge is to defamiliarize the arbitrary sign whose peculiar properties are so natural to us because they seem to pervade everything that is in any way human and anything else about which humans can hope to know. That you can feel *tsupu* without knowing Quichua makes language appear strange. It reveals that not all the signs with which we traffic are symbols and that those nonsymbolic signs can in important ways break out of bounded symbolic contexts like language. This explains not only why we can come to feel *tsupu* without speaking Quichua but also why Hilario can communicate with a nonsymbolic being. Indeed, the startled monkey's jump, and the entire ecosystem that sustains her, constitutes a web of semiosis of which the distinctive semiosis of her human hunters is just one particular kind of thread.

To summarize: signs are not exclusively human affairs. All living beings sign. We humans are therefore at home with the multitude of semiotic life. Our exceptional status is not the walled compound we thought we once inhabited. An anthropology that focuses on the relations we humans have with nonhuman beings forces us to step beyond the human. In the process it makes what we've taken to be the human condition—namely, the paradoxical, and "provincialized," fact that our nature is to live immersed in the "unnatural" worlds we construct—appear a little strange. Learning how to appreciate this is an important goal of an anthropology beyond the human.

THE FEELING OF RADICAL SEPARATION

The Amazon's many layers of life amplify and make apparent these greater than human webs of semiosis. Allowing its forests to think their ways through us can help us appreciate how we too are always, in some way or another, embedded in such webs and how we might do conceptual work with this fact.

This is what draws me to this place. But I've also learned something from attending to those times when I've felt cut off from these broader semiotic webs that extend beyond the symbolic. Here I reflect on such an experience that I had on one of the many bus trips I made from Quito to the Amazon region. I relay the feeling of what happened on this trip, not as a personal indulgence, but because I think it reveals a specific quality of symbolic modes of thinking—the propensity that symbolic thought has to jump out of the broader semiotic field from which it emerges, separating us, in the process, from the world around us. As such, this experience can also teach us something about how to understand the relation that symbolic thought has to the other kinds of thought in the world with which it is continuous and from which it emerges. In this sense, this reflection on my experience is also part of a broader critique, developed in the following two sections, of the dualistic assumptions at the base of so many of our analytical frameworks. I explore this experience of becoming dual, of feeling ripped out of a broader semiotic environment, that I had on a trip down to el Oriente, Ecuador's Amazonian region east of the Andes, by means of a narrative detour. Apart from serving as a bit of a respite from the conceptual work done in this chapter, I hope it will give some sense of the way in which Ávila itself is embedded in a landscape with a history. For this trip traces the trajectories of many other trips, and all of these catch this place up in so many kinds of webs.

The past few days had been unusually rainy on the eastern slopes of the Andes, and the main road leading down to the lowlands had been intermittently washed out. Joined by my cousin Vanessa, who was in Ecuador visiting relatives, I boarded a bus headed for the Oriente. With the exception of a group of Spanish tourists occupying the back rows, the bus was filled with locals who lived along the route or in Tena, the capital of Napo Province and the bus's final destination. This was a trip I had made many times by now, and it was our plan to take this bus along its route over the high cordillera east of Quito that divides the Amazonian watershed from the inter-Andean valley and then to follow this down through the village of Papallacta, the site of a pre-Hispanic cloud forest settlement situated along one of the major trade routes through which highland and lowland products flowed (I refer you to figure 1 on page 4). Today Papallacta is an important pumping station for Amazonian resources such as crude oil, which since the 1970s has transformed the country's economy and opened up the Oriente for development, and, more recently, drinking water for Quito tapped from the vast watershed east of the Andes. Nestled in a mountain chain that still experiences frequent geological activity, it is also the site of some very popular hot springs. Papallacta is, like many of the other cloud forest towns we would pass on our route, now mainly inhabited by highland settlers. The road is carved out of the precipitous gorges of the Quijos River valley, which it follows through what was the stronghold of the pre-Hispanic and early colonial alliance of Quijos chiefdoms. The ancestors of the Ávila Runa formed part of this alliance. Farmers regularly expose thousand-year-old residential terraces as they clear the steep forested slopes to create pastures. The route continues along the trajectory of the foot trails that until the 1960s connected Ávila and other lowland Runa villages like it, by means of an arduous eight-day journey, to Quito. We would take this road through the town of Baeza, which, along with Ávila and Archidona, was the first Spanish settlement founded in the Upper Amazon. Baeza was almost sacked in the same regionally coordinated 1578 indigenous uprising—sparked by the shamanic vision of a cow-god—that completely destroyed Ávila and left virtually all its Spanish inhabitants dead. Today's Baeza bears little resemblance to that historical town—having been relocated a few kilometers away following a large earthquake in 1987. Just before Baeza there is a fork in the road. One branch heads northeast toward the town of Lago Agrio. This was the first major center of oil extraction in Ecuador, and its name is a literal translation of Sour Lake, the site where oil was first discovered in Texas (and the birthplace of Texaco). The other branch, the one we would take, follows an older route to the town of Tena. In the 1950s Tena represented the boundary between civilization and the "savage" heathens (the Huaorani) to the east. Now it is a quaint town. After winding through steep and unstable terrain we would cross the Cosanga River where 150 years ago the Italian explorer Gaetano Osculati was abandoned by his Runa porters and forced to spend several miserable nights alone fending off jaguars (Osculati 1990). After this crossing there would be a final climb through the Huacamayos Cordillera, which is the last range to be traversed before dropping down to the warm valleys that lead to Archidona and Tena. On a clear day one can catch from here the shimmering reflections off the metal roofs in Archidona down below, as well as the road that goes from Tena to Puerto Napo, where it cuts a swath of red earth in the steep grade of a hill. Puerto Napo is the long abandoned "port" on the Napo River (indicated by a little anchor in figure 1), which flows into the Amazon. It had the misfortune of being situated just upstream from a dangerous whirlpool. If there are no clouds one can also see the sugar cone peak of the Sumaco Volcano on whose foothills Ávila sits. An area of close to 200,000 hectares making up the peak and many of its slopes is protected as a biosphere reserve. This reserve, in turn, is surrounded by a much larger area, which is designated as national forest. Ávila territory forms a border with this vast expanse on its western boundary.

Once out of the mountains the air becomes warmer and heavier as we pass little hamlets settled by lowland Runa. Finally, at another fork an hour before arriving at Tena, we would hop off to wait for a second bus that works its way along this decidedly more local and personal route. On this tertiary road a bus driver might stop to broker a deal on a few boxes of the tart *naranjilla* fruits used to make breakfast juice throughout Ecuador.[17] Or he might be persuaded to wait a few minutes for a regular passenger. This is a relatively new road, having been completed in the aftermath of the 1987 earthquake with the not entirely disinterested help of the U.S. Army Corps of Engineers. It winds through the foothills that circle Sumaco Volcano before heading out across the Amazonian plain at Loreto. It ends at the town of Coca at the confluence of the Coca and Napo Rivers. Coca, like Tena, but several decades later, also served as a frontier outpost of the Ecuadorian state as its control expanded deeper into this region. This road cuts through what used to be the hunting territories of the Runa villages of Cotapino, Loreto, Ávila, and San José, which, along with a handful of "white"-owned estates, or haciendas, and a Catholic mission in Loreto, were the only settlements in this area before the 1980s. Today large portions of these hunting territories are occupied by outsiders— either fellow Runa from the more densely populated Archidona region (whom people in Ávila refer to as *boulu*, from *pueblo*, referring to the fact that they are more city-wise) or small-time farmers and merchants of coastal or highland origin who are often referred to as *colonos* (or *jahua llacta*, in Quichua; lit., "highlanders").

Right after crossing the immense steel panel bridge that traverses the Suno River, one of several such structures along this route donated by the U.S. Army, we would get off at Loreto, the parish seat and biggest town on the road. We would spend the night here at the Josephine mission run by Italian priests. The following day we would retrace our steps, either by foot or by pickup truck, back over the bridge and then along a dirt road that follows the Suno River through colonist farms and pastures until we hit the trail leading to Ávila. Roads in eastern Ecuador extend in fits and starts over many years. Their growth spurts usually coincide with local election campaigns. When

I first started visiting Ávila in 1992 there were only foot trails from Loreto, and it would take me the better part of a day to get to Hilario's house. On my most recent visit one could, on a dry day, get to the easternmost portion of Ávila territory by pickup truck.

This was the route we had hoped to traverse. In fact, we didn't make it to Loreto that day. Not too far after Papallacta we encountered the first of a series of landslides set off by the heavy rains. And while our bus, along with a growing string of trucks, tankers, buses, and cars, waited for this to be cleared we became trapped by another landslide behind us.

This is steep, unstable, and dangerous terrain. The landslides reawakened in me a jumble of disturbing images from a decade of traveling this road: a snake frantically tracing figure eights in an immense mudflow that had washed over the road moments before we had gotten there; a steel bridge buckled in half like a crushed soda can by a slurry of rocks let loose as the mountain above it came down; a cliff splattered with yellow paint, the only sign left of the delivery truck that had careened into the ravine the night before. But landslides mostly cause delays. Those that can't quickly be cleared become sites for "*trasbordos*," an arrangement whereby oncoming buses that can no longer reach their destinations exchange passengers before turning back.

On this day a trasbordo was out of the question. Traffic was backed up in both directions, and we were trapped by a series of landslides scattered over a distance of several kilometers. The mountain above was starting to fall on us. At one point a rock crashed down onto our roof. I was scared.

No one else, however, seemed to think we were in danger. Perhaps out of sheer nerve, fatalism, or the need, above anything else, to complete the trip, neither the driver nor his assistant ever lost his cool. To a certain extent I could understand this. It was the tourists that baffled me. These middle-aged Spanish women had booked one of the tours that visit the rain forests and indigenous villages along the Napo River. As I worried, these women were joking and laughing. At one point one even got off the bus and walked ahead a few cars to a supply truck off of which she bought ham and bread and proceeded to make sandwiches for her group.

The incongruity between the tourists' nonchalance and my sense of danger provoked in me a strange feeling. As my constant what-ifs became increasingly distant from the carefree chattering tourists, what at first began as a diffuse sense of unease soon morphed into a sense of profound alienation.

This discrepancy between my perception of the world and that of those around me sundered me from the world and those living in it. All I was left with were my own thoughts of future dangers spinning themselves out of control. And then something more disturbing happened. Because I sensed that my thoughts were out of joint with those around me, I soon began to doubt their connection to what I had always trusted to be there for me: my own living body, the body that would otherwise give a home to my thoughts and locate this home in a world whose palpable reality I shared with others. I came, in other words, to feel a tenuous sense of existence without location—a sense of deracination that put into question my very being. For if the risks I was so sure of didn't exist—after all, no one else on that bus seemed frightened that the mountain would fall on us—then why should I trust my bodily connection to that world? Why should I trust "my" connection to "my" body? And if I didn't have a body what was "I"? Was I even alive? Thinking like this, my thoughts ran wild.

This feeling of radical doubt, the feeling of being cut off from my body and a world whose existence I no longer trusted, didn't go away when several hours later the landslides were cleared and we were able to get through. Nor did it subside when we finally got to Tena (it was too late to make it to Loreto that night). Not even in the relative comfort of my old haunt the hotel El Dorado did I manage to feel much better. This simple but cozy family-run inn used to be my stopping point when I was doing research in Runa communities on the Napo River.[18] It was owned by *don* Salazar, a veteran—with the scar to prove it—of Ecuador's short war with Peru in which Ecuador lost a third of its territory and access to the Amazon River. The hotel's name, El Dorado, appropriately marks this loss by paying homage to that never quite attainable City of Gold that lies somewhere deep in the Amazon (see Slater 2002; see also chapters 5 and 6).

The next morning after a fitful night I was still out of sorts. I couldn't stop imagining different dangerous scenarios, and I still felt cut off from my body and from those around me. Of course I pretended I wasn't feeling any of this. Trying at least to act normal, and in the process compounding my private anxiety by failing to give it a social existence, I took my cousin for a short walk along the banks of the Misahuallí River, which cuts the town of Tena in half. Within a few minutes I spotted a tanager feeding in the shrubs at the scruffy edges of town where molding cinder blocks meet polished river cobbles. I had brought along my binoculars and managed, after some searching, to locate the bird. I rolled the focusing knob and the moment that bird's thick black beak became sharp I experienced a sudden shift. My sense of separation simply dissolved. And, like the tanager coming into focus, I snapped back into the world of life.

There is a name for what I felt on that trip to the Oriente: anxiety. After reading *Constructing Panic* (1995), a remarkable account, written by the late psychologist Lisa Capps and the linguistic anthropologist Elinor Ochs, of one woman's lifelong struggles with anxiety, I've come to an understanding of this condition as revealing something important about the specific qualities of symbolic thought. Here is how Meg, the woman they write about, experiences the suffocating weight of all of the future possibles opened up by the symbolic imagination.

> Sometimes I get to the end of the day and feel exhausted by all of the "what if that had happened" and "what if this happens." And then I realize that I've been sitting on the sofa—that it's just me and my own thoughts driving me crazy. (Capps and Ochs 1995: 25)

Capps and Ochs describe Meg as "desperate" to "experience the reality that she attributes to normal people" (25). Meg feels "severed from an awareness of herself and her environment as familiar and knowable" (31). She senses that her experience does not fit with what, according to others, "happened" (24), and she thus has no one with whom to share a common image of the world, or a set of assumptions about how it works. Furthermore, she can't seem to ground herself in any specific place. Meg often uses the construction, "here I am," to express her existential predicament, but a crucial element is missing: "she is telling her interlocutors that she exists, but not where in particular she is located" (64).

The title *Constructing Panic* is intended by the authors to refer to how Meg discursively constructs her experience of panic—their assumption being that "the stories people tell construct who they are and how they view the world" (8). But I think the title reveals something deeper about panic. It is precisely the constructive quality of symbolic thought, the fact that symbolic thought can create so many virtual worlds, that makes anxiety possible. It is not just that Meg constructs her experience of panic linguistically, socially, culturally, in other words, symbolically, rather that panic itself is a symptom of symbolic construction run wild.

Reading Capps and Ochs's discussion of Meg's experience of panic, and thinking about it semiotically, I think I have come to an understanding of what happened on that trip to the Oriente, the factors that produced panic in me, and those that led to its dissipation. As with Meg, who locates her first experiences of anxiety in situations in which her legitimate fears were not socially recognized (31), my anxiety emerged as I was confronted with the disconnect between my well-founded fear and the carefree attitudes of the tourists on the bus.

Symbolic thought run wild can create minds radically separate from the indexical grounding their bodies might otherwise provide. Our bodies, like all of life, are the products of semiosis. Our sensory experiences, even our most basic cellular and metabolic processes, are mediated by representational— though not necessarily symbolic—relations (see chapter 2). But symbolic thought run wild can make us experience "ourselves" as set apart from every-thing: our social contexts, the environments in which we live, and ultimately even our desires and dreams. We become displaced to such an extent that we come to question the indexical ties that would otherwise ground this special kind of symbolic thinking in "our" bodies, bodies that are themselves indexi-cally grounded in the worlds beyond them: *I think therefore I doubt that I am.*

How is this possible? And why is it that we don't all live in a constant state of skeptical panic? That my sense of anxious alienation dissipated the moment the bird came into sharp focus provides some insights into the conditions under which symbolic thought can become so radically separate from the world, as well as those under which it can fall back into place. I do not, by any means, wish to romanticize tropical nature or privilege anyone's connection to it. This sort of regrounding can happen anywhere. Nonetheless, sighting that tanager in the bush at the messy edge of town taught me something about how immersion in this particularly dense ecology amplifies and makes visible a larger semiotic field beyond that which is exceptionally human, one in which we are all—usually—emplaced. Seeing that tanager made me sane by allowing me to situate the feeling of radical separation within something broader. It resituated me in a larger world "beyond" the human. My mind could return to being part of a larger mind. My thoughts about the world could once again become part of the thoughts of the world. An anthropology beyond the human strives to grasp the importance of these sorts of connections while appreciat-ing why we humans are so apt to lose sight of them.

NOVELTY OUT OF CONTINUITY

Thinking about panic in this way has led me to question more broadly how best to theorize the separation that symbolic thought creates. We tend to assume that because something like the symbolic is exceptionally human and thus novel (at least as far as earthly life is concerned) it must also be radically separate from that from which it comes. This is the Durkheimian legacy we inherit: social facts have their own kind of novel reality, which can only be understood in terms of other such social facts and not in terms of anything—be it psychological, biological, or physical—prior to them (see Durkheim 1972: 69–73). But the sense of radical separation that I experienced is psychically untenable—even life negating in some sense. And this leads me to suspect that there is something the matter with any analytical approach that would take such a separation as its starting point.

If, as I claim, our distinctively human thoughts stand in continuity with the forest's thoughts insofar as both are in some way or other the products of the semiosis that is intrinsic to life (see chapter 2), then an anthropology beyond the human must find a way to account for the distinctive qualities of human thought without losing sight of its relation to these more pervasive semiotic logics. Accounting conceptually for the relation this novel dynamic has to that from which it comes can help us better understand the relationship between what we take to be distinctively human and that which lies beyond us. In this regard I want to think here about what panic, and especially its resolution, has taught me. To do so I draw on a series of Amazonian examples to trace the ways in which iconic, indexical, and symbolic processes are nested within each other. Symbols depend on indices for their being and indices depend on icons. This allows us to appreciate what makes each of these unique without losing sight of how they also stand in a relation of continuity with each other.

Following Deacon (1997), I begin with a counterintuitive example at the very margins of semiosis. Consider the cryptically camouflaged Amazonian insect known as the walking stick in English because its elongated torso looks so much like a twig. Its Quichua name is *shanga*. Entomologists call it, appropriately, a phasmid—as in phantom—placing it in the order Phasmida and the family Phasmidae. This name is fitting. What makes these creatures so distinctive is their lack of distinction: they disappear like a phantom into the background. How did they come to be so phantasmic? The evolution of such creatures reveals important things about some of the "phantomlike" logical properties of semiosis that can, in turn, help us understand some of the counterintuitive properties of life "itself"—properties that are amplified in the Amazon and Runa ways of living there. For this reason, I will return to this example throughout the book. Here I want to focus on it with an eye to understanding how the different semiotic modalities—the iconic, the indexical, the symbolic—have their own unique properties at the same time that they stand in a relation of nested continuity to each other.

How did walking sticks come to be so invisible, so phantomlike? That such a phasmid looks like a twig does not depend on anyone noticing this resemblance—our usual understanding of how likeness works. Rather, its likeness is the product of the fact that the ancestors of its potential predators did not notice its ancestors. These potential predators failed to notice the differences between these ancestors and actual twigs. Over evolutionary time those lineages of walking sticks that were least noticed survived. Thanks to all the proto–walking sticks that were noticed—and eaten—because they differed from their environments walking sticks came to be more like the world of twigs around them.[19]

How walking sticks came to be so invisible reveals important properties of iconicity. Iconicity, the most basic kind of sign process, is highly counterintuitive because it involves a process by which two things are not distinguished. We tend to think of icons as signs that point to the similarities among things we know to be different. We know, for example, that the iconic stick figure of the man on the bathroom door resembles but is not the same as the person who might walk through that door. But there is something deeper about iconicity that is missed when we focus on this sort of example. Semiosis does not begin with the recognition of any intrinsic similarity or difference. Rather, it begins with not noticing difference. It begins with indistinction. For this reason iconicity occupies a space at the very margins of semiosis (for there is nothing semiotic about never noticing anything at all). It marks the beginning and end of thought. With icons new interpretants—subsequent signs that would further specify something about their objects—are no longer produced (Deacon 1997: 76, 77); with icons thought is at rest. Understanding something, however provisional that understanding may be, involves an icon. It involves a thought that is like its object. It involves an image that is a likeness of that object. For this reason all semiosis ultimately relies on the transformation of more complex signs into icons (Peirce CP 2.278).

Signs, of course, provide information. They tell us something new. They tell us about a difference. That is their reason for being. Semiosis must then involve something other than likeness. It must also involve a semiotic logic that points to something else—a logic that is indexical. How do the semiotic logics of likeness and difference relate to each other? Again, following Deacon (1997), consider the following schematic explanation of how that woolly monkey that Hilario and Lucio were trying to frighten out of her hidden canopy perch might learn to interpret a crashing palm as a sign of danger.[20] The thundering crash she heard would iconically call to mind past experiences of similar crashes. These past experiences of crashing sounds share with each other additional similarities, such as their co-occurrence with something dangerous—say, a branch breaking or a predator approaching. The monkey would in addition iconically link these past dangers to each other. That the sound made by a crashing tree might indicate danger is, then, the product of, on the one hand, iconic associations of loud noises with other loud noises, and, on the other, iconic associations of dangerous events with other dangerous events. That these two sets of iconic associations are repeatedly linked to each other encourages the current experience of a sudden loud noise to be seen as linked to them. But now this association is also something more than a likeness. It impels the monkey to "guess" that the crash must be linked to something other than itself, something different. Just as a wind vane, as an index, is interpreted as pointing to something other than itself, namely, the direction in which the wind is blowing, so this loud noise is interpreted as pointing to something more than just a noise; it points to something dangerous.

Indexicality, then, involves something more than iconicity. And yet it emerges as a result of a complex hierarchical set of associations among icons. The logical relationship between icons and indices is unidirectional. Indices are the products of a special layered relation among icons but not the other way around. Indexical reference, such as that involved in the monkey's take on the crashing tree, is a higher-order product of a special relationship among three icons: crashes bring to mind other crashes; dangers associated with such crashes bring to mind other such associations; and these, in turn, are associated with the current crash. Because of this special configuration of icons the current crash now points to something not immediately present: a danger. In this way an index emerges from iconic associations. This special relationship among icons results in a form of reference with unique properties that derive from but are not shared with the iconic associational logics with which they are continuous. Indices provide information; they tell us something new about something not immediately present.

Symbols, of course, also provide information. How they do so is both continuous with and different from indices. Just as indices are the product of relations among icons and exhibit unique properties with respect to these more fundamental signs, symbols are the product of relations among indices and have their own unique properties. This relationship also goes only in one direction. Symbols are built from a complex layered interaction among indices, but indices do not require symbols.

A word, such as *chorongo*, one of the Ávila names for woolly monkey, is a symbol par excellence. Although it can serve an indexical function—pointing to something (or, more appropriately, someone)—it does so indirectly, by virtue of its relation to other words. That is, the relation that such a word has to an object is primarily the result of the conventional relation it has acquired to other words and not just a function of the correlation between sign and object (as with an index). Just as we can think of indexical reference as the product of a special configuration of iconic relations, we can think of symbolic reference as the product of a special configuration of indexical ones. What is the relationship of indices to symbols? Imagine learning Quichua. A word such as *chorongo* is relatively easy to learn. One can learn that it refers to what in English is called a woolly monkey quite quickly. As such, it isn't really functioning symbolically. The pointing relationship between this "word" and the monkey is primarily indexical. The commands that dogs learn are very much like this. A dog can come to associate a "word" like *sit* with a behavior. As such, "sit" functions indexically. The dog can understand "sit" without understanding it symbolically. But there is a limit to how far we can go toward learning human language by memorizing words and what they point to; there are just too many individual sign-object relationships to keep track of. Furthermore, rote memorization of sign-object correlations misses the logic of language. Take a somewhat more complex word like *causanguichu*, which I discussed earlier in this chapter. Non-Quichua speakers can quickly learn that it is a greeting (uttered only in certain social contexts), but getting a sense of what and how it means requires us to understand how it relates to other words and even smaller units of language.

Words like *chorongo*, *sit*, or *causanguichu* do of course refer to things in the world, but in symbolic reference the indexical relation of word to object becomes subordinate to the indexical relation of word to word in a system of such words. When we learn a foreign language or when infants acquire language for the first time there is a shift away from using linguistic signs as indices to appreciating them in their broader symbolic contexts. Deacon (1997) describes one experimental setting where such a shift is particularly apparent. He discusses a long-term lab experiment in which chimps, already adept in their everyday lives at interpreting signs indexically, were trained to replace this interpretive strategy with a symbolic one.[21]

First, the chimps in the experiment had to interpret certain sign vehicles (in this case keyboard keys with certain shapes on them) as indices of certain objects or acts (such as particular food items or actions). Next, such sign vehicles had to be seen as indexically connected to each other in a systematic way. The final, and most difficult and most important, step involved an interpretive shift whereby objects were no longer picked out in a direct fashion by the individual indexical signs but instead came to be picked out indirectly, by virtue of the ways in which the signs representing them related to each other and the ways in which these sign relations then mapped onto how the objects themselves were to be thought to relate to each other. The mapping between these two levels of indexical associations (those linking objects to objects and those linking signs to signs) is iconic (Deacon 1997: 79–92). It involves not noticing the individual indexical associations by which signs can pick out objects in order to see a more encompassing likeness between the relations that link a system of signs and those that link a set of objects.

I am now in a position to account for the sense of separation—which I experienced as panic on the bus ride I described earlier—that the symbolic creates. I can now do so with regard to the more basic forms of reference to which it relates and with which it is continuous.

The symbolic is a prime example of a kind of dynamic that Deacon calls "emergent." For Deacon, an emergent dynamic is one in which particular configurations of constraints on possibility result in unprecedented properties at a higher level. Crucially, however, something that is emergent is never cut off from that from which it came and within which it is nested because it still depends on these more basic levels for its properties (Deacon 2006). Before considering symbolic reference as emergent with respect to other semiotic modalities it is useful to think about how emergence works in the nonhuman world.

Deacon recognizes a series of nested emergent thresholds. An important one is self-organization. Self-organization involves the spontaneous generation, maintenance and propagation of form under the right circumstances. Although relatively ephemeral and rare, self-organization is nonetheless found in the nonliving world. Examples of self-organizing emergent dynamics include the circular whirlpools that sometimes form in Amazonian rivers, or the geometric lattices of crystals or snowflakes. Self-organizing dynamics are more regular and more constrained than the physical entropic dynamics—such as those involved, for example, in the spontaneous flow of heat from a warmer to a colder part of a room—from which they emerge and on which they depend. Entities that exhibit self-organization, such as crystals, snowflakes, or whirlpools, are not alive. Nor, despite their name, do they involve a self.

Life, by contrast, is a subsequent emergent threshold nested within self-organization. Living dynamics, as represented by even the most basic organisms, selectively "remember" their own specific self-organizing configurations, which are differentially retained in the maintenance of what can now be understood as a self—a form that is reconstituted and propagated over the generations in ways that exhibit increasingly better fits to the worlds around it. Living dynamics, as I explore in greater detail in the following chapter, are constitutively semiotic. The semiosis of life is iconic and indexical. Symbolic reference, that which makes humans unique, is an emergent dynamic that is nested within this broader semiosis of life from which it stems and on which it depends.

Self-organizing dynamics are distinct from the physical processes from which they emerge and with which they are continuous, and within which they are nested. Living dynamics have a similar relation to the self-organizing dynamics from which they, in turn, emerge, and the same can be said for the relation that symbolic semiosis has to the broader iconic and indexical semiotic processes of life from which it emerges (Deacon 1997: 73).[22] Emergent dynamics, then, are directional both in a logical and in an ontological sense. That is, a world characterized by self-organization need not include life, and a living world need not include symbolic semiosis. But a living world must also be a self-organizing one, and a symbolic world must be nested within the semiosis of life.

I can now return to the emergent properties of symbolic representation. This form of representation is emergent with respect to iconic and indexical reference in the sense that, as with other emergent dynamics, the systemic structure of relationships among symbols is not prefigured in the antecedent modes of reference (Deacon 1997: 99). Like other emergent dynamics symbols have unique properties. The fact that symbols achieve their referential power by virtue of the systemic relations they have to each other means that, as opposed to indices, they can retain referential stability even in the absence of their objects of reference. This is what confers on symbols their unique characteristics. It is what allows symbolic reference to be not only about the here and now, but about the "what if." In the realm of the symbolic, the separation from materiality and energy can be so great and the causal links so convoluted that reference acquires a veritable freedom. And this is what has led to treating it as if it were radically separate from the world (see also Peirce CP 6.101).

Yet, like other emergent dynamics, such as the vortex of a whirlpool formed in a river's current, symbolic reference is also closely tied to the more basic dynamics out of which it grows. This is true in the way that symbols are constructed as well as in the way in which they are interpreted. Symbols are the outcome of a special relationship among indices, which in turn are outcomes of a special relationship that links icons in a particular way. And symbolic interpretation works via pairings of sets of indexical relations, which are ultimately interpreted by recognizing the iconicity between them: all thought ends with an icon. Symbolic reference, then, is ultimately the product of a series of highly convoluted systemic relations among icons. And yet it has properties that are unique when compared to iconic and indexical modalities. Symbolic reference does not exclude these other kinds of sign relations. Symbolic systems such as language can, and regularly do, incorporate relatively iconic signs, as in the case of "words" like *tsupu*, and they are also completely dependent on iconicity at a variety of levels as well as on all sorts of pointing relationships among signs and between systems of signs and the things they represent. Symbolic reference, finally, like all semiosis, is also ultimately dependent on the more fundamental material, energetic, and self-organizing processes from which it emerges.

Thinking of symbolic reference as emergent can help us understand how, via symbols, reference can become increasingly separated from the world but without ever fully losing the potential to be susceptible to the patterns, habits, forms, and events of the world.

Seeing symbolic reference and by extension human language and culture as emergent follows in the spirit of Peirce's critique of dualistic attempts to separate (human) mind from (nonhuman) matter—an approach that he acerbically characterized as "the philosophy which performs its analyses with an axe, leaving as the ultimate elements, unrelated chunks of being" (CP 7.570). An emergentist approach can provide a theoretical and empirical account of how the symbolic is in continuity with matter at the same time that it can come to be a novel causal locus of possibility. This continuity allows us to recognize how something so unique and separate is also never fully cut off from the rest of the world. This gets at something important about how an anthropology beyond the human seeks to situate that which is distinctive to humans in the broader world from which it emerges.

Panic and its dissipation reveal these properties of symbolic semiosis. They point both to the real dangers of unfettered symbolic thought and to how such thought can be regrounded. Watching birds regrounded my thoughts, and by extension my emerging self, by re-creating the semiotic environment in which symbolic reference is itself nested. Through the artifice of my binoculars I became indexically aligned with a bird, thanks to the fact that I was able to appreciate its image now coming into sharp focus right there in front of me. This event reimmersed me in something that Meg, on her sofa, alone with her thoughts, was not so readily able to find: a knowable (and shareable) environment, and the assurance, for the moment, of some sort of existence, tangibly located in a here and now that extended beyond me but of which I too could come to be a part.

Panic provides us with intimations of what radical dualism might feel like, and why for us humans dualism seems so compelling. In tracing its untenable effects panic also provides its own visceral critique of dualism and the skepticism that so often accompanies it. In panic's dissolution we can also get a sense for how a particular human propensity for dualism is dissolved into something else. One might say that dualism, wherever it is found, is a way of seeing emergent novelty as if it were severed from that from which it emerged.

EMERGENT REALS

By watching birds on the banks of the river that morning in Tena I certainly got out of my head in the colloquial sense, but what was I stepping into? Although the more basic semiotic modes of engagement involved in that activity quite literally brought me back to my senses and in the process regrounded me in a world beyond myself—beyond my mind, beyond convention, beyond the human—this experience has led me to ask what kind of world is this that lies out there beyond the symbolic? In other words, this experience, understood in the context of the anthropology beyond the human that I seek here to develop, forces me to rethink what we mean by the "real."

We generally think of the real as that which exists. The palm tree that came crashing down in the forest is real; the shorn branches and crushed plants left in the wake of its fall are proof of its awesome facticity. But a restricted characterization of the real as something that happened—out there and law-bound—can't account for spontaneity, or life's tendency for growth. Nor can it account for the semiosis shared by the living—a semiosis that emerges from and ultimately grounds us humans in the world of life. Furthermore, such a characterization would dualistically reinscribe all possibility in that separate chunk of being we delimit as the human mind with no intimation of how that mind, its semiosis and its creativity, could have emerged from or otherwise be related to anything else.

Peirce was quite concerned with this problem of how to imagine a more capacious real that is more true to a naturalistic, nondualist understanding of the universe and, throughout his career, strove to situate his entire philosophical project—including his semiotics—within a special kind of realism that could encompass actual existence within a broader framework that would account for its relationship to spontaneity, growth, and the life of signs in human and nonhuman worlds. I turn here to a brief exposition of his framework because it provides a vision of the real that can encompass living minds and nonliving matter, as well as the many processes through which the former emerged from the latter.

According to Peirce there are three aspects of the real of which we can become aware (CP 1.23–26). The element of the real that is easiest for us to comprehend is what Peirce called "secondness." The crashing palm is a quintessential second. Secondness refers to otherness, change, events, resistance, and facts. Seconds are "brutal" (CP 1.419). They "shock" (CP 1.336) us out of our habitual ways of imagining how things are. They force us to "think otherwise than we have been thinking" (CP 1.336).

Peirce's realism also encompasses something he called "firstness." Firsts are "mere may-bes, not necessarily realized." They involve the special kind of reality of a spontaneity, a quality, or a possibility (CP 1.304), in its "own suchness" (CP 1.424), regardless of its relation to anything else. One day out in the forest Hilario and I came across a bunch of wild passion fruits that had been knocked down by a troop of monkeys feeding up above. We took a break from our trek to snack on the monkeys' leftovers. As I cracked open the fruit, I caught, just for an instant, a pungent whiff of cinnamon. By the time I brought the fruit to my mouth it was gone. The experience of the fleeting smell, in and of itself, without attention to where it came from, what it is like, or to what it connects, approaches firstness.

Thirdness, finally, is that aspect of Peirce's realism that is the most important to the argument in this book. Drawing inspiration from the medieval Scholastics, Peirce insisted that "generals are real." That is, habits, regularities, patterns, relationality, future possibilities, and purposes—what he called thirds— have an eventual efficacy, and they can originate and manifest themselves in worlds outside of human minds (CP 1.409). The world is characterized by "the tendency of all things to take habits" (CP 6.101): the general tendency in the universe toward an increase in entropy is a habit; the less common tendency toward increases in regularity, exhibited in self-organizing processes such as the formation of circular whirlpools in a river or crystal lattice structures, is also a habit; and life, with its ability to predict and harness such regularities and, in the process, create an increasing array of novel kinds of regularities, amplifies this tendency toward habit taking. This tendency is what makes the world potentially predictable and what makes life as a semiotic process, which is ultimately inferential,[23] possible. For it is only because the world has some semblance of regularity that it can be represented. Signs are habits about habits. Tropical forests with their many layers of coevolved life-forms amplify this tendency toward habit taking to an extreme.

All processes that involve mediation exhibit thirdness. Accordingly, all sign processes exhibit thirdness because they serve as a third term that mediates between "something" and some sort of "someone" in some way. However, it is important to stress that for Peirce, although all signs are thirds, not all thirds are signs.[24] Generality, the tendency toward habit, is not a feature that is imposed on the world by a semiotic mind. It is out there. The thirdness in the world is the condition for semiosis, it is not something that semiosis "brings" to the world.

For Peirce everything exhibits, to some degree or other, firstness, secondness, and thirdness (CP 1.286, 6.323). Different kinds of sign processes amplify certain aspects of each of these to the neglect of others. Although all signs are intrinsically triadic, in that they all represent something to a someone, different kinds of signs attend more toward either firstness, secondness, or thirdness.

Icons, as thirds, are relative firsts in that they mediate by the fact that they possess the same qualities as their objects regardless of their relation to anything else. This is why Quichua imagistic "words" like *tsupu* cannot be negated or inflected. There is a way in which they are just qualities in their "own suchness." Indices, as thirds, are relative seconds because they mediate by being affected by their objects. The crashing palm startled the monkey. Symbols, as thirds, by contrast, are doubly triadic because they mediate by reference to something general—an emerging habit. They mean by virtue of the relationship they have to the conventional and abstract system of symbols—a system of habits—that will come to interpret them. This is why understanding *causanguichu* requires a familiarity with Quichua as a whole. The symbolic is a habit about a habit that, to a degree unprecedented elsewhere on this planet, begets other habits.

Our thoughts are like the world because we are of the world.[25] Thought (of any kind) is a highly convoluted habit that has emerged out of, and is continuous with, the tendency in the world toward habit taking. In this manner Peirce's special kind of realism can allow us to begin to envision an anthropology that can be about the world in ways that recognize but also go beyond the limits of human-specific ways of knowing. Rethinking semiosis is the place from which to begin such an endeavor.

It is through this expanded vision of the real that we can consider what it was that I was getting out of when that bird came into focus through the glass of my binoculars, and what it was in that process that I stepped into. As Capps and Ochs astutely point out, what is so disturbing about panic is the feeling of being out of sync with others. We come to be alone with thoughts that become increasingly cut off from the broader field of habits that gave rise to them. In other words, there is always the danger that symbolic thought's unmatched ability to create habit can pull us out of the habits in which we are inserted.

But the living mind is not uprooted in this way. Thoughts that grow and are alive are always about something in the world, even if that something is a potential future effect. Part of the generality of thought—its thirdness—is that it is not just located in a single stable self. Rather, it is constitutive of an emerging one distributed over multiple bodies:

> Man is not whole as long as he is single[;] . . . he is essentially a possible member of society. Especially, one man's experience is nothing, if it stands alone. If he sees what others cannot, we call it hallucination. It is not "my" experience, but "our" experience that has to be thought of; and this "us" has indefinite possibilities. (Peirce CP 5.402)

This "us" is a general.

And panic disrupts this general. With panic there is a collapse of the triadic relation linking my habit-making mind to other habit-making minds vis-à-vis our ability to share the experience of the habits of the world that we discover.

The solipsistic enfolding of an increasingly private mind onto itself results in something terrifying: the implosion of the self. In panic the self becomes a monadic "first" severed from the rest of the world; a "possible member of society" whose only capability is to doubt the existence of any of what Haraway (2003) calls its more "fleshly" connections to the world. The result, in sum, is a skeptical Cartesian *cogito*: a fixed "I (only) think (symbolically) therefore I (doubt that I) am" instead of a growing, hopeful, and emergent "us" with all its "indefinite possibilities."[26]

This triadic alignment that results in an emergent "us" is achieved indexically and iconically. Consider Lucio's running commentary after he shot the woolly monkey that had been scared out of her treetop perch by the palm tree that Hilario felled:

> there
> right there
> there
> what's gonna happen?
> there, it's curled up in a ball
> all wounded[27]

Hilario, whose eyesight is not as good as Lucio's, wasn't immediately able to see the monkey up in the tree. Whispering, he asked his son, "Where?" And as the monkey suddenly began to move Lucio rapidly responded, "Look! look! look! look!"

The imperative "look!" (Quichua *"ricui!"*) functions here as an index to orient Hilario's gaze along the path of the monkey's movement across the length of the branch. As such it aligns Hilario and Lucio vis-à-vis the monkey in the tree. In addition, Lucio's rhythmic repetition of the imperative iconically captures the pace of the monkey's movement along the branch. Through this image that Hilario can also come to share, Lucio can "directly communicate" his experience of seeing the wounded monkey moving through the canopy, regardless of whether his father actually managed to see her.

It is precisely this sort of iconic and indexical alignment that brought me back into the world the moment that tanager came into focus in my binoculars. That crisp image of the bird sitting right there in those shrubs grounded me again in a shareable real. This is so even though icons and indices do not provide us with any immediate purchase on the world. All signs involve mediation, and all of our experiences are semiotically mediated. There is no bodily, inner, or other kind of experience or thought that is unmediated (see Peirce CP 8.332). Furthermore, there is nothing intrinsically objective about this real tanager feeding on a real riverbank plant. For this animal and its shrubby perch—like me—are semiotic creatures through and through. They are the results of representation. They are outcomes of an evolutionary process of ever-increasing alignment with those proliferating webs of habits that constitute tropical life. Such habits are real, regardless of whether or not I can appreciate them. By acquiring a feel for some of these habits, as I did with that tanager on the river's edge that morning, I can potentially become aligned with a broader "us" thanks to the way others can share this experience with me.

Like our thoughts and minds, birds and plants are emergent reals. Life-forms, as they represent and amplify the habits of the world, create new habits, and their interactions with other organisms create even more habits. Life, then, proliferates habits. Tropical forests, with their high biomass, unparalleled species diversity, and intricate coevolutionary interactions, exhibit this tendency toward habit taking to an unusual degree. For people like the Ávila Runa, who are intimately involved with the forest through hunting and other subsistence activities, being able to predict these habits is of the utmost importance.

So much of what draws me to the Amazon is the ways in which one kind of third (the habits of the world) are represented by another kind of third (the human and nonhuman semiotic selves who live in and constitute this world) in such a way that more kinds of thirds can "flourish" (see Haraway 2008). Life proliferates habits. Tropical life amplifies this to an extreme, and the Runa and others who are immersed in this biological world can amplify this even further.

GROWTH

Being alive—being in the flow of life—involves aligning ourselves with an ever-increasing array of emerging habits. But being alive is more than being in habit. The lively flourishing of that semiotic dynamic whose source and outcome is what I call self is also a product of disruption and shock. As opposed to inanimate matter, which Peirce characterized as "mind whose habits have become fixed so as to lose the powers of forming them and losing them," mind (or self) "has acquired in a remarkable degree a habit of taking and laying aside habits" (CP 6.101).

This habit of selectively discarding certain other habits results in the emergence of higher-order habits. In other words, growth requires learning something about the habits around us, and yet this often involves a disruption of our habituated expectations of what the world is like. When the pig that Maxi shot plunged—tsupu—into the river, as wounded pigs are known to do, Maxi assumed that he had gotten his quarry. He was wrong:

> foolishly, "it's gonna die," I'm thinking
> when
> it suddenly ran off [28]

Maxi's feeling of bewilderment occasioned by the supposedly dead peccary suddenly jumping up and running off reveals something of what Haraway (1999: 184) calls "a sense of the world's independent sense of humor." And it is in such moments of "shock" that the habits of the world make themselves manifest. That is, we don't usually notice the habits we in-habit. It is only when the world's habits clash with our expectations that the world in its otherness, and its existent actuality as something other than what we currently are, is revealed. The challenge that follows this disruption is to grow. The challenge is to create a new habit that will encompass this foreign habit and, in the process, to remake ourselves, however momentarily, anew, as one with the world around us.

Living in and from the tropical forest requires an ability to make sense of the many layers of its habits. This is sometimes accomplished by recognizing those elements that appear to disrupt them. On another walk in the forest with Hilario and his son Lucio we came across a small bird of prey, known in English as the hook-billed kite,[29] perched in the branches of a small tree. Lucio shot at it but missed. Frightened, the bird flew off in a strange manner. Rather than fly rapidly through the understory, as raptors are expected to do, it lumbered off quite slowly. As he pointed in the direction in which it went Lucio remarked:

> it just went off slowly
> tca tca tca tca
> there[30]

Tca tca tca tca. Throughout the day Lucio repeated this sonic image of wings flapping slowly, hesitantly, and somewhat awkwardly.[31] The kite's cumbersome flight caught Lucio's attention. It disrupted the expectation that raptors should exhibit swift and powerful flight. Similarly the ornithologists Hilty and Brown (1986: 91) describe the hook-billed kite as having unusually "broad lanky wings" and being "rather sedentary and sluggish." Compared to other raptors that exhibit swifter flight, this bird is anomalous. It disrupts our assumptions about raptors, and this is why its habits are interesting.

Another example: upon returning home one morning from a hunt Hilario pulled out from his net bag an epiphytic cactus (*Discocactus amazonicus*) dotted with purple flowers. He called it *viñarina panga* or *viñari panga*, because, as he explained, "*pangamanda viñarin*," "it grows out of its leaves." It has no particular use, although, like other succulent epiphytes such as orchids, he thought that the macerated stem might make a good poultice to apply to cuts. But because the leaves of this plant appear to grow out of other leaves, Hilario found this plant strange. The name "viñari panga" gets at a botanical habit that extends deep into the evolutionary past. Leaves do not grow out of other leaves. They can only grow out of the meristematic tissue located in buds on twigs, stems, and branches. The ancestral group within the cacti, from which *D. amazonicus* is derived, originally lost its laminar photosynthetic leaves and developed succulent rounded photosynthetic stems. Those flattened green structures that grow out of each other in *D. amazonicus* are therefore not true leaves. They are actually stems that function as leaves and for this reason they can grow out of each other. These leaflike stems appear to put into question the habit that leaves sprout from stems. This is what makes them interesting.

WHOLES PRECEDE PARTS

In semiosis, as in biology, wholes precede parts; similarity precedes difference (see Bateson 2002: 159). Thoughts and lives both begin as wholes—albeit ones that can be extremely vague and underspecified. A single-celled embryo, however simple and undifferentiated, is just as whole as the multicellular organism into which it will develop. An icon, however rudimentary its likeness, insofar as it is taken as a likeness, imperfectly captures the object of its similarity as a whole. It is only in the realm of the machine that the differentiated part comes first and the assembled whole second.[32] Semiosis and life, by contrast, begin whole.

An image, then, is a semiotic whole, but as such it can be a very rough approximation of the habits it represents. One afternoon while drinking manioc beer at Ascencio's house we heard Sandra, Ascencio's daughter, cry out from her garden some way off, "A snake! Come kill it!"[33] Ascencio's son Oswaldo rushed out, and I followed close behind. Although the creature in question turned out to be an inoffensive whipsnake,[34] Oswaldo killed it anyway with a blow from the broad side of his machete and then severed and buried its head.[35] As we walked back to the house Oswaldo pointed out a little stump that I had just stumbled on and noted that he had seen me stumble on the very same stump the day before on our return along that path after a long day out hunting with his father and brother-in-law in the steep forested slopes west of Ávila.

On those walks with Oswaldo back to the house my ambulatory habits had only imperfectly matched the habits of the world. Because of fatigue or mild inebriation (the first time I had stumbled on that stump we had hiked more than ten hours over very steep terrain and I was exhausted, the second time I had just finished off several big bowls of manioc beer) I simply failed to interpret some of the features of the path as salient. I acted as if there were no obstacles. I could get away with this because my regular gait was an interpretive habit—an image of the path—that was good enough for the challenge at hand. Given the conditions that we faced it didn't really matter if the way I walked didn't perfectly match the features of the path. If, however, we had been running, or if I had been burdened by a heavy load, or if it had been raining heavily, or if I had been a little bit more tipsy, that lack of fit may well have become amplified, and instead of slightly stumbling I might well have tripped and fallen.

My tipsy or fatigued representation of the forest path was so rudimentary that I failed to notice its differences. Until Oswaldo pointed it out to me I never noticed the stump, or that I had stumbled on it—twice! My stumbling had become its own fixed habit. By virtue of the regularity my imperfect walking habit had assumed—so regular that I could repeatedly kick the same stump on successive days—it became visible to Oswaldo as its own anomalous habit. And yet, however imperfect its match to the path, my manner of walking was good enough. It got me home.

But there was something lost in that "good enough" habituated automatization. Perhaps that day walking back to Ascencio's house, I had become, for a moment, more like matter—"mind whose habits had become fixed"—and less a learning and yearning, living and growing self.

Unexpected events, such as the sudden appearance of a stump across our path—when we manage to notice it—or Maxi's peccary suddenly reviving can disrupt our assumptions of how the world is. And it is this very disruption, the breakdown of old habits and the rebuilding of new ones, that constitutes our feeling of being alive and in the world. The world is revealed to us, not by the fact that we come to have habits, but in the moments when, forced to abandon our old habits, we come to take up new ones. This is where we can catch glimpses—however mediated—of the emergent real to which we also contribute.

THE OPEN WHOLE

Recognizing how semiosis is something broader than the symbolic can allow us to see the ways we come to inhabit an ever-emerging world beyond the human. An anthropology beyond the human aims to reach beyond the confines of that one habit—the symbolic—that makes us the exceptional kinds of beings that we believe we are. The goal is not to minimize the unique effects this habit has but only to show some of the different ways in which the whole that is the symbolic is open to those many other habits that can and do proliferate in the world that extends beyond us. The goal, in short, is to regain a sense of the ways in which we are open wholes.

This world beyond the human, to which we are open, is more than something "out there" because the real is more than that which exists. Accordingly, an anthropology beyond the human seeks a slight displacement of our temporal focus to look beyond the here and now of actuality. It must, of course, look back to constraints, contingencies, contexts, and conditions of possibility. But the lives of signs, and of the selves that come to interpret them, are not just located in the present, or in the past. They partake in a mode of being that extends into the future possible as well. Accordingly, this anthropology beyond the human aims to attend to the prospective reality of these sorts of generals as well as to their eventual effects in a future present.

If our subject, the human, is an open whole, so too should be our method. The particular semiotic properties that make humans open to the world beyond the human are the same ones that can allow anthropology to explore this with ethnographic and analytical precision. The realm of the symbolic is an open whole because it is sustained by, and ultimately cashed out in, a broader, different kind of whole. That broader whole is an image. As Marilyn Strathern once said to me, paraphrasing Roy Wagner, "You can't have half an image." The symbolic is one particular human-specific way to come to feel an image. All thought begins and ends with an image. All thoughts are wholes, however long the paths that will bring them there may be.[36]

This anthropology, like semiosis and life, does not start with difference, otherness, or incommensurability. Nor does it start with intrinsic likeness. It begins with the likeness of thought-at-rest—the likeness of not yet noticing those eventual differences that might come to disrupt it. Likenesses, such as tsupu, are special kinds of open wholes. An icon is, on the one hand, monadic, closed unto itself, regardless of anything else. It is like its object whether or not that object exists. I feel tsupu whether or not you do. And yet, insofar as it stands for something else, it is an opening as well. An icon has the "capacity of revealing unexpected truth": "by direct observation of it other truths concerning its object can be discovered" (Peirce CP 2.279). Peirce's example is an algebraic formula: because the terms to the left of the equals sign are iconic of those to the right we can learn something more about the latter by considering the former. That which is to the left is a whole. It captures that which is to its right in its totality. And yet in the process it is also able to suggest, "in a very precise way, new aspects of supposed states of things" (CP 2.281). This is possible, thanks to the general way it stands for this totality. Signs stand for objects "not in all respects but in reference to a sort of idea" (CP 2.228). This idea, however vague, is a whole.

Attending to the revelatory power of images suggests a way to practice an anthropology that can relate ethnographic particulars to something broader. The inordinate emphasis on iconicity in lowland Quichua amplifies and makes apparent certain general properties of language and the relation that language has to that which lies beyond it, just as panic exaggerates and therefore makes apparent other properties. These amplifications or exaggerations can function as images that can reveal something general about their objects. Such generals are real despite the fact that they lack the concreteness of the specific or the fixed normativity of those putative universals that anthropology rightly rejects. It is to such general reals that an anthropology beyond the human can gesture. It does so, however, in a particularly worldly way. It grounds itself in the mundane strivings and stumblings that emerge in the ethnographic moment, with a view to how such contingent everydays make apparent something about general problems.

My hope is that this anthropology can open itself to some of the new and unexpected habits just coming into being that might catch it up. By opening itself to novelty, images, and feelings, it seeks the freshness of firstness in its subject and method. I ask you to feel tsupu for yourself, and this is something I cannot force upon you. But it is also an anthropology of secondness in that it hopes to register how it is surprised by the effects of such spontaneities as they come to make a difference in a messy world that is the emergent product of all the ways in which its motley inhabitants engage with and attempt to make sense of each other. Finally, this is an anthropology of the general, for it aims to recognize those opportunities where an *us* that exceeds the limits of individual bodies, species, and even concrete existence can come to extend beyond the present. This *us*—and the hopeful worlds it beckons us to imagine and realize—is an open whole.

TWO

The Living Thought

Funes not only remembered every leaf on every tree of every wood, but even every one of the times he had perceived or imagined it. . . . I suspect, however, that he wasn't very capable of thinking. Thinking is forgetting differences.

—Jorge Luis Borges, *Funes el Memorioso*

Harvesting fish poison roots[1] in the woody thickets that used to be their gardens, Amériga and Luisa were within earshot when it happened. Back at home, as they talked with Delia over bowls of manioc beer, Luisa imitated how through the brush she had heard the family's dogs—Pucaña, or Red Face, their favorite; Cuqui, her aging companion; and Huiqui—barking excitedly, "'*hua' hua' hua' hua' hua' hua' hua' hua' hua*,'" the way they do when they're following game. Then she heard them barking, "'*ya ya ya ya*,'" poised to attack. But then something very disturbing happened. The dogs started yelping, "'*aya—i aya—i aya—i*,'" indicating that now they had been attacked and were in great pain.

"And that," Luisa remarked, "was it. They just fell silent."[2]

chun
silence

How could things have changed so suddenly? For the women, the answer turned on imagining how the dogs understood, or, more accurately, failed to understand, the world around them. Reflecting on the first two series of barks, Luisa remarked, "That's what they'd do if they came across something big." That's what they would do, that is, if they came across a big game animal. "'Was it a deer they were barking at?'" Luisa remembered asking herself. That would make sense. Just a few days before, the dogs had tracked down, attacked, and killed a deer. And we were still eating the meat.

But what creature might look to the dogs like prey but then turn on them? The women concluded that there was only one possible explanation; the dogs must have confused a mountain lion with a red brocket deer. Both have tawny coats and are approximately the same size. Luisa tried to imagine what they were thinking: "It looks like a deer, let's bite it!"

Delia concisely summed up their frustration with the dogs' confusion: "So so stupid." Amériga elaborated: "How is it that they didn't know? How is it that they could even think [of barking], 'yau yau yau,' as if they were going to attack it?"

What each bark meant was clear, for these barks are part of an exhaustive lexicon of canine vocalizations that people in Ávila feel they know. What was less obvious was what, from the dogs' perspectives, prompted them to bark in those ways. To imagine that the dogs might fail to discriminate between a mountain lion and a deer and to trace out the tragic consequences of that confusion—the dogs just saw something big and tawny and attacked it—required thinking beyond what in particular the dogs did, to how it was that what they did was motivated by how they came to understand the world around them. The conversation began to revolve around the question of how dogs think.

This chapter develops the claim that all living beings, and not just humans, think, and explores another closely related claim, that all thoughts are alive. It is about "the living thought."[3] What does it mean to think? What does it mean to be alive? Why are these two questions related, and how does our approach to them, especially when seen in terms of the challenges of relating to other kinds of beings, change our understanding of relationality and "the human"?

If thoughts are alive and if that which lives thinks, then perhaps the living world is enchanted. What I mean is that the world beyond the human is not a meaningless one made meaningful by humans.[4] Rather, mean-ings— means-ends relations, strivings, purposes, telos, intentions, functions and significance—emerge in a world of living thoughts beyond the human in ways that are not fully exhausted by our all-too-human attempts to define and control these.[5] More precisely, the forests around Ávila are animate. That is, these forests house other emergent loci of mean-ings, ones that do not necessarily revolve around, or originate from, humans. This is what I'm getting at when I say that forests think. It is to an examination of such thoughts that this anthropology beyond the human now turns.

If thoughts exist beyond the human, then we humans are not the only selves in this world. We, in short, are not the only kinds of we. Animism, the attribution of enchantment to these other-than-human loci, is more than a belief, an embodied practice, or a foil for our critiques of Western mechanistic representations of nature, although it is also all of these as well. We should not, then, just ask how some humans come to represent other beings or entities as animate; we also need to consider more broadly what is it about these that make them animate.

People in Ávila, if they are to successfully penetrate the relational logics that create, connect, and sustain the beings of the forest, must in some way recognize this basic animacy. Runa animism, then, is a way of attending to living thoughts in the world that amplifies and reveals important properties of lives and thoughts. It is a form of thinking about the world that grows out of a specially situated intimate engagement with thoughts-in-the-world in ways that make some of their distinctive attributes visible. Paying attention to these engagements with the living thoughts of the world can help us think anthropology differently. It can help us imagine a set of conceptual tools we can use to attend to the ways in which our lives are shaped by how we live in a world that extends beyond the human.

Dogs, for example, are selves because they think. Counterintuitively, however, proof that they think is that they, in Delia's words, can be "so so stupid"— so indifferent, so dumb. That the dogs in the forest were considered capable of confusing a mountain lion with a deer suggests an important question: How is it that indifference, confusion, and forgetting are so central to the lives of thoughts and the selves that come to house them? The strange and productive power of confusion in the living thought challenges some of our basic assumptions about the roles that difference and otherness, on the one hand, and identity, on the other, play in social theory. This can help us rethink relationality in ways that can take us beyond our tendency to apply our assumptions about the logic of linguistic relationality to all the many possible ways in which selves might relate.

NONHUMAN SELVES

The women certainly felt they were able to interpret the dogs' barks, but that's not what makes them recognize their dogs as selves. What makes their dogs selves is that their barks were manifestations of their interpretations of the world around them. And how those dogs interpreted the world around them, as the women were amply aware, matters vitally. We humans, then, are not the only ones who interpret the world. "Aboutness"—representation, intention, and purpose in their most basic forms—is an intrinsic structuring feature of living dynamics in the biological world. Life is inherently semiotic.[6]

This intrinsically semiotic characteristic applies to all biological processes. Take for example the following evolutionary adaptation: the elongated snout and tongue of the giant anteater. The giant anteater, or *tamanuhua*, as it is known in Ávila, can be deadly if cornered. One Ávila man was almost killed by one during my time there (see chapter 6), and even jaguars are said to keep well away from them (see chapter 3). The giant anteater is also ethereal. I caught a fleeting glimpse of one off in the distance in the forest as Hilario, Lucio, and I were resting on a log on a ridge above the Suno River late one afternoon. Its image still impresses itself on me today: the silhouette of a tapered head, a stocky body, and an enormous splayed fan of a tail around whose hairs the late afternoon sun's rays passed.

Giant anteaters feed exclusively on ants. They do so by inserting their elongated snouts into ant colony tunnels. The specific shape of the anteater's snout and tongue captures certain features of its environment, namely, the shape of ant tunnels. This evolutionary adaptation is a sign to the extent that it is interpreted (in a very bodily way, for there is no consciousness or reflection here) by a subsequent generation with respect to what this sign is about (i.e., the shape of ant tunnels). This interpretation, in turn, is manifested in the development of the subsequent organism's body in a way that incorporates these adaptations. This body (with its adaptations) functions as a new sign representing these features of the environment, insofar as it, in turn, will be interpreted as such by another subsequent generation of anteaters in the eventual development of that generation's body.

Anteater snouts over the generations have come to represent with increasing accuracy something about the geometry of ant colonies because those lineages of "protoanteaters" whose snouts and tongues less accurately captured relevant environmental features (e.g., the shapes of ant tunnels) did not survive as well. Relative to these protoanteaters, then, today's living anteaters have come to exhibit comparatively increasing "fittedness" (Deacon 2012) to these environmental features. They are more nuanced and exhaustive representations of it.[7] It is in this sense that the logic of evolutionary adaptation is a semiotic one.

Life, then, is a sign process. Any dynamic in which "something . . . stands to somebody, for something in some respect or capacity," as Peirce's (CP 2.228) definition of a sign has it, would be alive. Elongated snouts and tongues *stand to* a future anteater (a "somebody") *for* something about the architecture of an ant colony. One of Peirce's most important contributions to semiotics is to look beyond the classical dyadic understanding of signs as something that stands for something else. Instead, he insisted, we should recognize a crucial third variable as an irreducible component of semiosis: signs stand for something in relation to a "somebody" (Colapietro 1989: 4). As the giant anteater illustrates, this "somebody"—or a self, as I prefer to call it—is not necessarily human, and it need not involve symbolic reference, subjectivity, the sense of interiority, consciousness, or the awareness we often associate with representation for it to count as such (see Deacon 2012: 465–66).

Furthermore, selfhood is not limited just to animals with brains. Plants are also selves. Nor is it coterminous with a physically bounded organism. That is, selfhood can be distributed over bodies (a seminar, a crowd, or an ant colony can act as a self), or it can be one of many other selves within a body (individual cells have a kind of minimal selfhood).

Self is both the origin and the product of an interpretive process; it is a waypoint in semiosis (see chapter 1). A self does not stand outside the semiotic dynamic as "Nature," evolution, watchmaker, homuncular vital spirit, or (human) observer. Rather, selfhood emerges from within this semiotic dynamic as the outcome of a process that produces a new sign that interprets a prior one. It is for this reason that it is appropriate to consider nonhuman organisms as selves and biotic life as a sign process, albeit one that is often highly embodied and nonsymbolic.

MEMORY AND ABSENCE

The giant anteater as a self is a form that selectively "remembers" its own form. That is, a subsequent generation is a likeness of a previous one. It is an iconic representation of its ancestor. But at the same time as such an anteater is a likeness of its forebear (and is thus a sort of memory of it) it also differs from it. For this anteater, with its snout and tongue, can potentially be a relatively more detailed representation of the world around it, insofar (in this case) as its snout, when compared to that of its ancestor, better fits ant tunnels. In sum, the way this anteater remembers or re-presents the generations that came before it is "selective." This is so, in part, thanks to those past protoanteater selves whose snouts didn't "fit" their environments as well and who were thus, in a sense, forgotten.

This play of remembering and forgetting is both unique and central to life; any lineage of living organism—plant or animal—will exhibit this characteristic. Contrast this with, say, a snowflake. Although the particular form that a given snowflake takes is a historically contingent product of the interaction with its environment as it falls to the ground (and this is why we think of snowflakes as exhibiting a sort of individuality; no two are alike) the particular form a snowflake takes is never selectively remembered. That is, once it melts its form will have no bearing on the form that any subsequent snowflake will take as it begins to fall to the ground.

Living beings differ from snowflakes because life is intrinsically semiotic, and semiosis is always for a self. The form an individual anteater takes comes to represent, for a future instantiation of itself, the environment its lineage has come to fit over evolutionary time. Anteater lineages selectively remember their previous fits to their environments; snowflakes don't.

A self, then, is the outcome of a process, unique to life, of maintaining and perpetuating an individual form, a form that, as it is iterated over the generations, grows to fit the world around it at the same time that it comes to exhibit a certain circular closure that allows it to maintain its selfsame identity, which is forged with respect to that which it is not (Deacon 2012: 471); anteaters re-present previous representations of ant tunnels in their lineage, but they are not themselves ant tunnels. Insofar as it strives to maintain its form, such a self acts for itself. A self, then, whether "skin-bound" or more distributed, is the locus of what we can call agency (479–80).

Because a giant anteater is a sign, what it is—its particular configuration, the fact, for example, that it has an elongated, as opposed to some other shape of snout—cannot be understood without considering what it is about, namely, the relevant environment that it increasingly comes to fit through the dynamic I've just described. Therefore, although semiosis is embodied, it also always involves something more than bodies. It is about something absent: a semiotically mediated future environment, which is potentially like the environment to which the past generation fit (see chapter 1).

A living sign is a prediction of what Peirce calls a habit (see chapter 1). That is, it is an expectation of a regularity, something that has not yet come to exist but will likely come to be. Snouts are products of what they are not, namely, the possibility that there will be ant tunnels in the environment into which the snouted anteater will come to live. They are the products of an expectation— of a highly embodied "guess" at what the future will hold.

This is a result of another important absence. As I mentioned earlier, the snouts and the way they fit with the world around them are the result of all the previous wrong "guesses"—the previous generations whose snouts were less like that world of ant tunnels. Because the snouts of these protoanteaters didn't fit the geometry of ant tunnels quite as well as the snouts of others, their forms did not survive into the future.

This way in which selves strive to predict "absent" futures also manifests itself in the purported behavior of América's dogs. The dogs must have barked, the women imagined, at what they expected and trusted was a deer. More accurately, perhaps, they barked at something they saw as big and tawny. Unfortunately, however, mountain lions are also big and tawny. A semiotically mediated future—the possibility of attacking the perceived deer—came to affect the present. It influenced the dogs' decision—"so stupid" in hindsight—to chase the creature they thought was prey.

LIFE AND THOUGHT

A lineage of signs can potentially extend into the future as an emergent habit, insofar as each instantiation will interpret the previous one in a way that can, in turn, be interpreted by a future one. This applies equally to a biological organism, whose progeny may or may not survive into the future, as it does to this book, whose ideas may or may not be taken up in the thinking of a future reader (see Peirce CP 7.591). Such a process is what constitutes life. That is, any kind of life, be it human, biological, or even, someday, inorganic, will spontaneously exhibit this embodied, localized, representational, future-predicting dynamic that captures, amplifies, and proliferates the tendency toward habit taking in a future instantiation of itself. Another way of saying this is that any entity that stands as a locus of aboutness, within a lineage of such loci that can potentially extend into the future, can be said to be alive. The origins of life—any kind of life, anywhere in the universe—also necessarily marks the origins of semiosis and of self.

It also marks the origins of thought. Life-forms—human and nonhuman alike—because they are intrinsically semiotic, exhibit what Peirce calls a "'scientific' intelligence." By "scientific," he does not mean an intelligence that is human, conscious, or even rational but simply one that is "capable of learning by experience" (CP 2.227). Selves, as opposed to snowflakes, can learn by experience, which is another way of saying that, through the semiotic process I've been describing, they can grow. And this, in turn, is another way of saying that selves think. Such thinking need not happen in the time scale we chauvinistically call real time (see Dennett 1996: 61). It need not happen, that is, within the life of a single skin-bound organism. Biological lineages also think. They too, over the generations, can grow to learn by experience about the world around them, and as such they too demonstrate a "'scientific' intelligence." In sum, because life is semiotic and semiosis is alive, it makes sense to treat both lives and thoughts as "living thoughts." This deepened understanding of the close relationship between life, self, and thought is central to this anthropology beyond the human that I am developing here.

AN ECOLOGY OF SELVES

The semiotic quality of life—the fact that the forms that life takes are the product of how living selves represent the world around them—structures the tropical ecosystem. Although all life is semiotic, this semiotic quality is amplified and made more apparent in the tropical forest, with its unparalleled kinds and quantities of living selves. This is why I want to find ways to attend to how forests think; tropical forests amplify, and thus can make more apparent to us, the ways life thinks.

The worlds that selves represent are not just made up of things. They are also, in large part, made up of other semiotic selves. For this reason I have come to refer to the web of living thoughts in and around the forests of Ávila as an ecology of selves. This ecology of selves in and around Ávila includes the Runa as well as other humans who interact with them and the forest, and it holds in its configurations not only the many kinds of living beings of the forest but also, as I discuss toward the end of this book, the spirits and the dead that make us the living beings that we are.

How different kinds of beings represent and are represented by other kinds of beings defines the patterning of life in the forests around Ávila. For example, once a year the colonies of leafcutter ants (*Atta* spp.)—whose presence is normally visible only in the long files of workers carrying to their nests snippets of vegetation they have culled from treetops—change their activity. Over the space of a few minutes, each widely dispersed colony simultaneously disgorges hundreds upon hundreds of plump winged reproductive ants and sends them flying into the early morning sky to mate with those from other colonies. This event poses, and indeed is structured by, a variety of challenges and opportunities. How do the ants, living in far-flung colonies, manage to coordinate their flights? How can predators tap into this rich but ephemeral cache? And what strategies do the ants use to avoid being eaten? These flying ants, overburdened with fat reserves, are a savory delicacy that people in Ávila, as well as many others who live in the Amazon, covet. Indicative of how much they are valued, they are known simply as *añangu*, ants. Toasted with salt they are a delicacy, and collected by the potful they are an important food source during the limited time they are available. How do people manage to predict the few minutes in each year when these will come out of their underground nests?

The problem of when the ants fly can tell us something about how the rain forest comes to be what it is: an emergent and expanding multilayered cacophonous web of mutually constitutive, living, and growing thoughts. Because in this part of the equatorial tropics there are no marked seasonal changes in sunlight or temperature, and no corresponding spring bloom, there is no one stable cue external to the interactions among forest beings that determines or predicts when ants will fly. The timing of this event is a product of the coordinated prediction of seasonal meteorological regularities as well as an orchestration among different, competing, and interpreting species.

According to people in Ávila the winged ants emerge in the calm that follows a period of heavy rains that includes thunder and lightning and the flooding of rivers. This stormy period brings to a close a relatively drier period that usually occurs around August. People try to predict the emergence of the ants by linking it to a variety of ecological signs associated with fruiting regimes, increases in insect populations, and changes in animal activity.[8] When the various indicators point to the fact that "ant season" (*añangu uras*) is at hand, people go to the various nests around their houses several times throughout the night to check for the telltale signs that the ants will soon take flight. These signs include the presence of guards clearing entrances of debris and sightings of a few slowly emerging and still somewhat lethargic winged ants.

People in Ávila are not the only ones interested in when these ants will fly. Other creatures, such as frogs, snakes, and small felines,[9] are attracted to the ants, as well as to those other animals that are attracted to the ants. They all watch the ants and watch those watching the ants for signs of when the ants will emerge from their nests.

Although the day of the flight is closely linked to meteorological patterns, and this seems to be how the ants coordinate their flights with those from other nests, the precise moment at which the flight will take place on that day is a response, sedimented over evolutionary time, to what it is that potential predators might, or might not, notice. It is no accident that the ants take flight just before daybreak (at exactly 5:10, when I've been able to time it). When they are in their nests the aggressive guards of the colony protect them from snakes, frogs, and other predators. Once they take flight, however, they are on their own, and they can fall prey to the lingering fruit-eating bats still out at twilight who attack them in midflight by biting off their greatly enlarged, fat-filled abdomens.

How bats see the world matters vitally to the flying ants. It is no accident that the ants take flight at the time they do. Although some lingering bats are still out, by this time they will only be active for twenty or thirty minutes longer. When the birds come out (not long after a six o'clock sunrise) most of the ants will have already dispersed, and some females will have already copulated and fallen to the ground to establish new colonies. The precise timing of the ant flight is an outcome of a semiotically structured ecology. The ants emerge at twilight—that blurry zone between night and day—when nocturnal and diurnal predators are least likely to notice them.

People attempt to enter some of the logic of the semiotic network that structures ant life in order to capture the ants during those few minutes in the year when they fly out of their nests. One night, as the ants were about to fly, Juanicu asked me for a cigarette so that he could blow tobacco smoke infused with the power of his "life breath" (*samai*) in order to send the impending rain clouds away. If it rained that evening the ants would not emerge. His wife, Olga, however, urged him not to ward off the rain clouds. She feared that their sons, who had gone to market in Loreto, would not return from town until the following day. They would be needed to harvest the ants that would be pouring out of the various nests around the house. To make sure the ants would not fly that night, she went out to all the nearby nests and stomped on them. This, she said, would keep the ants from coming out that evening.

On the night that Juanicu felt sure the ants would finally fly, he urged me, before I went out with his children in the middle of the night to check the nests, not to kick or step heavily around the nest. Then, shortly before five in the morning, at a distance of about four meters from the entrance of the nest closest to the house, Juanicu and I placed some lit kerosene lanterns as well as some of my candles and my flashlight. The winged ants are attracted to light and would be drawn to these sources. The lights were placed far enough away, however, so that the guards would not consider them threatening.

As the ants began to emerge Juanicu spoke only in whispers. Shortly after five o'clock we could hear a buzzing as the winged ants began to come out from the nest and fly off. Many of these were attracted by the light and came to us instead of flying to the sky. Juanicu then began to whistle like a siren alternating between two different pitches. This, he later explained, is something the flying ants understand as the call of their "mothers."[10] As the ants came to us, we singed off their wings with torches made of dry *lisan* leaves.[11] We were then easily able to place them into covered pots.[12]

The leafcutter ants are immersed in an ecology of selves that has shaped their very being; that they emerge just before dawn is an effect of the interpretive propensities of their major predators. People in Ávila also attempt to tap into the communicative universe of the ants and of the many creatures connected to them. Such a strategy has practical effects; people are able to gather vast quantities of ants based on them.

By treating ants as the intentional communicating selves they are, Juanicu was able to arrive at an understanding of the various associations that link ants to the other beings in the forest—an understanding that is surely never absolute but sufficient to accurately predict the few moments in the year when these ants will fly. He was also able to communicate directly with them, calling them to their deaths. In doing so he was, in effect, entering the logic of how forests think. This is possible because his (and our) thoughts are in important respects like those that structure the relations among those living thoughts that make the forest what it is: a dense, flourishing, ecology of selves.

SEMIOTIC DENSITY

The interrelations among so many different semiotic life-forms in this dense ecology of selves result in a relatively more nuanced and exhaustive overall representation of the surrounding environment when compared to the way life represents elsewhere on the planet. That is, the "thoughts" of a tropical forest come to represent the world in a relatively more detailed way. For example, a number of tropical tree species have evolved as specialists that grow only on white-sand soils. Tropical white-sand soils, as contrasted to tropical clay soils, are nutrient-poor, do not hold water well, and have characteristics such as high acidity that can slow plant growth. However, it is not the soil conditions in themselves that account for the fact that there are specialists that live on white-sand soils. Rather, the fact that there are such specialists is the result of their relation to another set of life-forms: plant-eating organisms, or herbivores (Marquis 2004: 619).

Because of the extremely poor conditions of these white-sand soils, plants have difficulty repairing themselves fast enough to sustain the levels of nutrient loss incurred by herbivory. Thus there is great selective pressure for plants living on such nutrient-poor soils to develop highly specialized toxic compounds and other defenses against herbivory (Marquis 2004: 620).

Interestingly, however, soil differences do not directly affect what kinds of plants can grow where. Fine, Mesones, and Coley (2004) have shown that when herbivores are experimentally removed from poor-soil plots and rich-soil species experimentally transplanted in the rich-soil species actually grow better than those adapted to poor soils.

So one could say that tropical plants come to represent something about their soil environments by virtue of their interactions with the herbivores that amplify the differences in soil conditions and thus make these differences important to plants. That is, differences in soil types wouldn't make a difference to the plants if it weren't for these other life-forms. This is why rich-soil plants, not burdened by the need to produce energetically costly toxins, grow better than poor-soil plants in poor-soils plots that have been kept free of herbivores experimentally.[13]

In temperate regions, where insectivorous herbivores are far fewer, there is very little specialization of plants to soil type even in areas where soil heterogeneity (i.e., the juxtaposition of nutrient rich and poor soils) is higher than in tropical regions (Fine 2004: 2). Another way to say this is that plants in the tropics, as opposed to those in temperate regions, come to form relatively more nuanced representations of the characteristics of their environments. They make more differentiations among soil types because of the ways they are caught up in a relatively denser web of living thoughts.

This herbivore-dependent amplification effect of soil differences does not stop with plants but continues to propagate through the ecology of selves. Tannin, for example, is a chemical defense that many Amazonian poor-soil plants have developed against herbivores. Because microorganisms cannot easily break down tannin-rich leaf litter, this compound leaches into rivers where it is toxic to fish and many other organisms. As a consequence, ecosystems associated with rivers that drain large expanses of white-sand soil are not able to support as much animal life (Janzen 1974), and historically this has had an important impact on humans living in the Amazon (Moran 1993). The various forms that all these ecologically related kinds of life take are not reducible to the characteristics of soil. I'm not making an argument for environmental determinism.[14] And yet this multispecies assemblage captures and amplifies something about the differences in soil conditions precisely as a function of the greater number of relations (relative to other ecosystems) among kinds of selves that exist in this ecology of selves.

RELATIONALITY

Selves, in short, are thoughts, and the modes by which such selves relate to one another stem from their constitutively semiotic nature and the particular associational logics this entails. Considering the logic by which these selves relate in this ecology of selves challenges us to rethink relationality—arguably our field's fundamental concern and central analytic (Strathern 1995).

If selves are thoughts and the logic through which they interact is semiotic, then relation is representation. That is, the logic that structures relations among selves is the same as that which structures relations among signs. This, in itself, is not a new idea. Whether or not we are explicit about it we already tend to think of relationality in terms of representation in the ways we theorize society and culture. But we do so based on our assumptions about how human symbolic representation works (see chapter 1). Like the words that exist in the conventional relational configurations that make up a language, the relata—be these ideas, roles, or institutions—that make up a culture or a society, do not precede the mutually constitutive relationships these relata have with one another in a system that necessarily comes to exhibit a certain closure by virtue of this fact.

Even posthuman relational concepts, such as Bruno Latour's "actant," the networks of actor-network theory, and Haraway's "constitutive intra-action" (Haraway 2008: 32, 33), rely on assumptions about relationality that stem from the special kinds of relational properties we find in human language. In fact, in some versions of actor-network theory the relational networks that connect humans and nonhuman entities are explicitly described as language-like (see Law and Mol 2008: 58).[15]

But representation, as I have been arguing, is something both broader than and different from what we expect given how our thinking about it has been linguistically colonized. Extending linguistic relationality to nonhumans narcissistically projects the human onto that which lies beyond it. And along with language comes a host of assumptions about systematicity, context, and difference, which stem from some of the distinctive properties of human symbolic reference and are not necessarily relevant to how living thoughts might more generally relate. In the process, other properties that might permit a more capacious view of relationality are obscured. My claim, in short, is that an anthropology beyond the human can rethink relationality by seeing it as semiotic but not always and necessarily languagelike.

Consider, in this regard, the relation between a wood tick and the mammals it parasitizes, a relation made classic by the early-twentieth-century ethologist Jakob von Uexküll (1982). Ticks, according to von Uexküll, perceive mammals, whose blood they suck, from the smell of butyric acid, warmth, and the ability to detect the bare patches of mammalian skin where they can burrow. According to him, their experiential world, or *umwelt* as he called it, is limited to just these three parameters (Uexküll 1982: 57, 72). For von Uexküll, and many of those who have picked up on his work, the tick's experiential world is closed and "poor," in the sense that the tick doesn't differentiate among many entities (see Agamben 2004). But I want to emphasize the productive power of this simplification that is central to living thoughts and to the relations that emerge among the selves that are the products of living thoughts. And I want to highlight the fact that its relational logic is semiotic but not distinctively symbolic.

Ticks do not distinguish among many kinds of mammals. It makes no difference to ticks that, for example, a dog might be wise to distinguish a predatory mountain lion from potential prey like the red brocket deer. The tick will confuse these two with each other and these with the dogs as well.

Ticks are also vectors for parasites, and because of the ways ticks fail to distinguish among mammals whose blood they indiscriminately suck, these parasites can pass from one species to another. This indiscrimination is a form of confusion, which of course has its limits. If the tick confused everything with everything else, there would be no thinking here and no life; confusion is only productive when it is constrained.

For the tick, one kind of mammal is, in Peircean terms, iconic of another. I want to highlight this view of iconism, which I introduced in the previous chapter, because it goes against our everyday understandings of the term. When we treat icons (signs that signify through similarity) we usually think of the ways in which we take them to be like some aspect of something else that we already know to be different. We do not, as I mentioned, confuse a stick figure depiction of a man posted on the door of a washroom with the person who might enter through that door. But I'm alluding here to a more fundamental—and often misunderstood—iconic property, one that underlies all semiosis. To the tick, mammals are equivalent, simply because the tick doesn't notice the differences among the beings it parasitizes.

This iconic confusion is productive. It creates "kinds." There emerges a general class of beings whose members are linked to each other because of the ways they are all noticed by ticks, who do not discriminate among them. This emergence of a general class matters to the beings involved. Because the tick confuses these warm-blooded beings, other parasites can travel among them (the "mammals") through the tick. This, in fact, is how Lyme disease is transmitted from deer to humans.

The world of living beings is neither just a continuum nor a collection of disparate singularities waiting to be grouped—according to social convention or innate propensity—by a human mind. It is true that categorization can be socioculturally specific and that it can lead to a form of conceptual violence in that it erases the uniqueness of those categorized. And it is also true that the power of human language lies in its ability to jump out of the local in ways that can result in an increased insensitivity to detail. Speaking of a Japanese insect collector, Hugh Raffles writes:

> After collecting for so many years, he now has *"mushi"* eye, bug eyes, and sees everything in nature from an insect's point of view. Each tree is its own world, each leaf is different. Insects taught him that general nouns like *insects, trees, leaves,* and especially *nature* destroy our sensitivity to detail. They make us conceptually as well as physically violent. "Oh, an insect," we say, seeing only the category, not the being itself. (2010: 345)

And yet seeing the world with "bug eyes" in many instances actually involves confusing what we might otherwise take as different entities, and this sort of confusion is neither exclusively human nor only destructive.

Borges's character Ireneo Funes, mentioned in this chapter's epigraph, was thrown by a wild horse and suffered a head injury, with the result that he could no longer forget anything. He became *"memorioso."* But living selves are precisely not like Funes, who was incapable of forgetting the distinctive features of "every leaf on every tree of every wood." This, as Borges points out, is not thinking. The life of thoughts depends on confusion—a kind of "forgetting" to notice difference. Generals, such as kinds and classes, emerge from and flourish in the world through a form of relating based on confusion. The real is not just the unique singularity, different from everything else. Generals are also real, and some generals emerge as a product of the relations among living thoughts beyond the human.

KNOWING WITHOUT KNOWING

How could Amériga, Delia, and Luisa presume to guess at what their dogs were thinking? More generally, how can we ever hope to know these other living selves with whom we relate? Even if we grant that nonhuman life-forms are selves, doesn't there exist, in Derrida's (2008: 30) words, such an "abyssal rupture" separating us from them that theirs might be better thought as an "existence that refuses to be conceptualized" (9)? Might these "absolute other[s]" (11) not be like Wittgenstein's lion; even if they could talk, who would understand them? Thomas Nagel's (1974) answer to the question he posed his fellow philosophers, *What is it like to be a bat?*, was decisive; although there is surely something that it is like to be a bat—that bats, in effect, have some kind of selfhood—we can never know it. We are just too different.

Granted, Amériga, Luisa, and Delia will never know with certainty what their dogs were thinking as they barked at that feline moments before it attacked them, but they could make some good guesses. What, then, might a theory of relating look like that started, not with the search for some secure knowledge of other beings, but with the sorts of provisional guesses that these women were forced to make about the guesses their dogs, in turn, might be making? Such a theory would not begin with what Haraway (2003: 49) calls "irreducible difference," nor would it take the refusal to be conceptualized, or its logical opposite, absolute understanding, as inhabitable endpoints.

Absolute otherness, irreducible difference, incommensurablity—these are taken to be the hurdles that our theories of relating must strive to overcome. That there exist differences that are radically inconceivable—differences that are so unimaginable that they are "incognizable" as Peirce (1992d: 24) critically calls them—implies an opposite: that knowability is based on intrinsic self-similarity. It implies that there exists such a thing as "being itself" in all its singularity, which we might comprehend if we could just adopt "bug eyes." These poles are taken to define how beings can relate to and know each other.

However, when we consider "the living thought," similarity and difference become interpretive positions (with potential future effects). They are not intrinsic characteristics that are immediately apparent. "All thought and knowledge," writes Peirce, "is by signs" (CP 8.332). That is, all thinking and knowing is mediated in some way.

This has important implications for understanding relating. There is no inherent difference between the associations of living thoughts that constitute the living thinking knowing self and those by which different kinds of selves might relate and thereby form associations. Further, because selves are loci of living thoughts—emergent ephemeral waypoints in a dynamic process—there is no unitary self. There is no one thing that one could "be": "[A] person is not absolutely an individual. His thoughts are what he is 'saying to himself,' that is, is saying to that other self that is just coming into life in the flow of time" (Peirce CP 5.421). Because all experiences and all thoughts, for all selves, are semiotically mediated, introspection, human-to-human intersubjectivity, and even trans-species sympathy and communication are not categorically different. They are all sign processes. For Peirce, the Cartesian *cogito*, the "I think," is not exclusively human, nor is it housed inside the mind, nor does it enjoy any exclusive or unmediated purchase on its most intimate object: the self that we commonly think of as the one doing our thinking.

Peirce illustrates this by asking us to imagine what red looks like to others. Far from being a private phenomenon, he argues, we can be pretty confident that we can have some sense of this. We can even have some idea of what this color is like to a blind person who has never seen red but who gathers from others that it resembles the sound of trumpets: "The fact that I can see a certain analogy, shows me not only that my feeling of redness is something like the feelings of the persons whom he had heard talk, but also his feeling of a trumpet's blare was very much like mine" (CP 1.314).[16] Peirce concludes by suggesting that self-knowledge is ultimately like these processes: "My metaphysical friend who asks whether we can ever enter into one another's feelings ... might just as well ask me whether I am sure that red looked to me yesterday as it does today" (CP 1.314). Introspection and intersubjectivity are semiotically mediated. We can only come to know ourselves and others through the medium of signs. It makes no difference whether that interpreting self is located in another kind of body or whether it is "that other self"—our own psychological one—"that is just coming into life in the flow of time," as one

FIGURE 5. What a hawk looks like to a parakeet. Photo by author.

sign is interpreted by a new one in that semiotic process by which thoughts, minds, and our very being qua self emerge.

Rather than make knowledge of selves impossible, this mediation is the basis for its possibility. Because there is no absolute "incognizable" there is also no absolute incommensurability. We can know something of how red might be experienced by a blind person, what it might be like to be a bat, or what those dogs might have been thinking moments before they were attacked, however mediated, provisional, fallible, and tenuous these understandings may be. Selves relate the way that thoughts relate: we are all living, growing thoughts.

A simple example illustrates this. The Runa make scarecrows, or more accurately "scare-parakeets," in order to scare white-eyed parakeets from their cornfields. They do so by binding together in a cross two flattened pieces of balsa wood of equal length. They paint these with red and black stripes using *achiote*[17] and charcoal, respectively. They also carve the top part to fashion a head and paint big eyes on it, and they sometimes insert the distinctively barred tail feathers of an actual raptor at the ends of the pieces of wood that will serve to represent the tail and the wings (see figure 5).

The elaborate fashion in which the Runa decorate this scarecrow is not an attempt to "realistically" represent a raptor from the human point of view. Rather, it constitutes an attempt to imagine what from the parakeet's perspective a raptor looks like. The scarecrow is an icon. It stands for a raptor by virtue of the likeness it has with the raptor for somebody—here, the parakeet. By virtue of stripes, big eyes, and actual tail feathers, the scarecrow captures something of what a raptor is like for a parakeet. This is why parakeets, but not humans, confuse these scarecrows with raptors. Proof of this is that these scarecrows successfully keep parakeets away and are thus made from year to year in Ávila. We can know something of what it is like to be a parakeet, and we know this by the effects that our guesses at how parakeets think can have on them.

ENCHANTMENT

It is very difficult from within our contemporary analytical frameworks to understand the biological world as made up of living thoughts. This, following Max Weber's (1948a, 1948b) diagnosis of the disenchantment of the modern world, is in part an effect of the spread of scientific rationalism. As we come to increasingly see the world in mechanistic terms we lose sight of the telos, the significance, the means-ends relations—in short, the mean-ings, as I call them, to highlight the close relationship between means and meanings—that were once recognized in the world. The world becomes disenchanted in the sense that ends are no longer to be found in the world. The world becomes literally meaningless. Ends become displaced to a human or spiritual realm that becomes ever smaller and more detached from the mundane world as this vision of science expands to encompass more domains.

If modern forms of knowledge and ways of manipulating the nonhuman world are characterized by an understanding of the world as mechanism, then disenchantment is an obvious consequence. Machines, as material objects, are means to achieve ends that are, by definition and design, external to them. When we contemplate a machine—say, a dishwasher—we bracket out the ends that are actually intrinsic to its being, namely, that it was built for some end by somebody. Applying this logic to the nonhuman living world, seeing nature as a machine, requires a similar bracketing and a subsequent ascription of ends to humans, gods, or Nature. Dualism is one result of this bracketing. Another is that we begin to lose sight of ends altogether. Disenchantment spreads into the realm of the human and the spiritual as we come to suspect that perhaps there simply are no ends and hence no meaning—anywhere.

But ends are not located somewhere outside the world but constantly flourishing in it. They are intrinsic to the realm of life. Living thoughts "guess" at and thus create futures to which they then shape themselves. Nor is the logic that structures the living world like that of a machine. Unlike machines, living thoughts emerge whole instead of being built from parts by someone bracketed out of the picture. If we attend to Runa engagements with other kinds of beings, as I aim to do here through this anthropology beyond the human, we can come to appreciate selves (both human and nonhuman) as waypoints in the lives of signs—loci of enchantment—and this can help us imagine a different sort of flourishing in this world beyond the human in which we live.

I'm making a claim here about some of the properties of life "itself." Although I recognize how something like life itself can be historically circumscribed—that certain concepts can only become thinkable in specific historical, social, or cultural contexts (Foucault 1970)—I want to reiterate something I discussed more fully in the first chapter. Language and the related discursive regimes that condition so much of our thought and action are not closed. Although we must of course be cautious about the ways in which language (and by extension, certain socially stabilized modes of thought and action) naturalizes categories of thought, we can venture to talk about something like life "itself" without being fully constrained by the language that carries this forth.

Nonhuman selves, then, have ontologically unique properties associated with their constitutively semiotic nature. And these are, to a certain extent, knowable to us. These properties differentiate selves from objects or artifacts. Treating nonhumans generically—indiscriminately lumping together things and beings—however, misses this. And this, to my mind, is the biggest shortcoming of STS, the dominant approach for expanding the social sciences to consider nonhumans.

STS brings nonhumans and humans into the same analytical framework through a form of reductionism that leaves concepts like agency and representation unexamined. As a consequence the distinctively human instantiations of these become stand-ins for all agency and representation. The result is a form of dualism in which humans and nonhumans acquire mixtures of thinglike and humanlike properties (see chapter 1).

Latour (1993, 2004), the main proponent of this approach, for example, attributes agency either to that which can be represented or to that which can resist our attempts at representation (see also Pickering 1999: 380–81). But these characteristics only capture, what, in Peircean terms would be called the secondness, that is the actuality or brute factuality, of the entity in question (see chapter 1)—for anything can potentially resist representation or be represented—and this simply reinstates the material/meaning divide STS tries to overcome. We still have, on the one hand, the material (now agentified), and, on the other, those humans (now made a little more obtuse and less certain of their omniscience) who represent or misrepresent things, as the case may be.

But resistance is not agency. Conflating resistance and agency blinds us to the kinds of agency that do in fact exist beyond the human. Because telos, representation, intentionality, and selfhood still need to be accounted for and because the way such processes emerge and operate beyond the human is not theorized, Latourian science studies is forced to fall back on humanlike forms of representation and intentionality as operative in the world beyond the human. These are then applied, if only metaphorically, to entities otherwise understood only in their secondness.

Substances, for example, undergo the "'sufferings'" of trials (Latour 1987: 88), and they sometimes emerge successfully as "heroes" (89). The piston of an engine is more reliable than a human operator, "since it is, via the cam, *directly interested*, so to speak, in the right timing of steam. Certainly it is more directly interested than any human being" (130; Latour's emphasis). And scientists use "a set of strategies to enlist and interest the human actors, and a second set to enlist and interest the non-human actors so as to hold the first" (132).

This approach to nonhuman agency overlooks the fact that some nonhumans, namely, those that are alive, are selves. As selves, they are not just represented, but they also represent. And they can do so without having to "speak." Nor do they need a "spokesperson" (Latour 2004: 62–70) because, as I discussed in chapter 1, representation exceeds the symbolic, and it therefore exceeds human speech.

Although we humans certainly represent nonhuman living beings in a number of culturally, historically, and linguistically distinct ways, and this surely has its effects, both for us and for those beings thus represented, we also live in worlds in which how these selves represent us can come to matter vitally. Accordingly, my concern is with exploring interactions, not with nonhumans generically—that is, treating objects, artifacts, and lives as equivalent entities—but with nonhuman living beings in terms of those distinctive characteristics that make them selves.

Selves, not things, qualify as agents. Resistance is not the same as agency. Nor, contra Bennett (2010), does materiality confer vitality. Selves are the product of a specific relational dynamic that involves absence, future, and growth, as well as the ability for confusion. And this emerges with and is unique to living thoughts.

ANIMISM

I want to return to the anecdote with which I began this book. Recall that when I was in the forest on a hunting trip I was told to make sure to sleep faceup. This way if a jaguar were to pass by he would see me as a being capable of looking back and would leave me alone. If I were to sleep facedown, I was warned, that potential passing jaguar might well treat me as prey and attack me. My point was that this anecdote forces us to recognize that how jaguars see us matters to us, and that if this is so, then anthropology cannot limit itself to asking how people see the world. I noted that by returning the feline's gaze, we allow jaguars the possibility of treating us as selves. If, by contrast, we were to look away, they would treat us as, and we may actually become, objects— literally, dead meat, aicha.

The linguist Émile Benveniste (1984) observes that the pronouns *I* and *you* position interlocutors intersubjectively through mutual address, and accordingly he considers these true "person" pronouns. By contrast, the third person is more accurately a "non-person" (Benveniste 1984: 221). It refers to something outside of the discursive interaction. If we extend this reasoning to trans-species encounters, then jaguars and humans, in this act of looking back at each other, would, in a sense, become persons to each other. And in the process, the Runa, in a way, would also become jaguars.

Indeed, as I mentioned in the introduction, the Ávila Runa are renowned—and feared—throughout lowland Runa communities for their ability to become shape-shifting were-jaguars. A person who is treated by a jaguar as prey may well become dead meat. By contrast, one who is treated by a jaguar as a predator becomes another predator. Predator and prey—puma and aicha—are the two kinds of beings that jaguars recognize. As with the tick, how jaguars represent other beings makes beings into kinds. And what kind of a being one thus becomes matters.

Puma in Quichua simply means "predator." For example, in Ávila the name for the crab-eating raccoon,[18] whose diet includes, among other things, crusta-ceans and mollusks, is *churu puma*, snail predator. Because the jaguar exempli-fies the quintessence of predation, it is simply known as puma. Runa who survive encounters with such predators are by definition, then, runa puma, or were-jaguars (the term *Runa* is not only an ethnonym; it also means "person" [see chapter 6]). One survives, then, by not being noticed as prey by a puma. But in the process one also becomes another kind of being, a puma. And this newfound status translates to other contexts and creates new possibilities.

Puma is a relational category—not, in this respect, unlike the pronouns *I* and *you* (see chapter 6). That we can become puma by returning a puma's gaze is a way of saying that we both are kinds of *Is*—that we both are kinds of persons. The Runa, like other Amazonians, treat jaguars and many other non-human beings as soul-possessing, signifying, intentional selves. They are (to use a recently resuscitated term) animists; for them, nonhumans are animate. They are persons.

Animism, as it is currently being theorized by people like Descola (2005) and Viveiros de Castro (1998), is quite different from its earlier social evolu-tionist and sometimes even racist incarnations, and it has provided an impor-tant foil for critiquing Western mechanistic representations of nature. And yet such critiques of the ways we in the "West" represent nature only asks how other humans come to treat nonhumans as animate. In this respect these approaches stand in continuity with such classical treatments of animism as

Lévy-Bruhl's *How Natives Think* (1926). The case of the jaguar troubles this project; if jaguars also represent us we cannot just ask how it is that some of us humans happen to represent them as doing so.

Animism, to my mind, gets at something more far reaching about the properties of the world, and this is why thinking with it is central to an anthropology beyond the human. It captures an animation that is emergent with life, hence my title, *How Forests Think*. Runa animism grows out of a need to interact with semiotic selves qua selves in all their diversity. It is grounded in an ontological fact: there exist other kinds of thinking selves beyond the human.

I recognize of course that those we call animists may well attribute animacy to all sorts of entities, such as stones, that I would not, according to the framework laid out here, consider living selves. If I were building an argument from within a particular animistic worldview, if I were routing all my argumentation through what, say, the Runa think, say, or do, this discrepancy might be a problem. But I don't. Part of my attempt to open anthropology to that which lies beyond the human involves finding ways to make general claims about the world. These claims don't necessarily line up with certain situated human viewpoints, like, say, those of animists, or those of biologists, or those of anthropologists.

How Forests Think, not *How Natives Think, about Forests* (cf. Sahlins 1995): if we limit our thinking to thinking through how other people think we will always end up circumscribing ontology by epistemology (chapter 1 suggests a solution to this problem). I am making here a general claim about selfhood. This general claim—which is not exactly an ethnographic one in the sense that it is not circumscribed by an ethnographic context, even though it is suggested, explored, and defended, in part, ethnographically—is that living beings are loci of selfhood. I make this claim empirically. It grows out of my attention to Runa relations with nonhuman beings as these reveal themselves ethnographically. These relations amplify certain properties of the world, and this amplification can infect and affect our thinking about the world.

One might say that the animal person is the model of the universe for animists, whereas for us it is the machine. Ontologically speaking, each has its own truth: animals are persons, and there are things about the world that do resemble partible machines (which is a reason why reductionist science is so successful). But my goal here is not to say which one is right or to point out where each fails but rather to see how certain kinds of engagements, based on certain presuppositions that themselves grow out of those engagements, amplify unexpected and real properties of the world that we can harness to think beyond the human as we know it.

Runa animism is pragmatically oriented. The challenge for the Runa, as people who engage intimately with the beings of the forest in order, in large part, to eat them, is to find ways to enter this vast ecology of selves to harness some of its plenty. This requires being attuned to the unexpected affinities we share with other selves while at the same time recognizing the differences that distinguish the many kinds of selves that people the forest.

PERSPECTIVISM

Like many Amazonians, people in Ávila approach this through a way of understanding others that Viveiros de Castro (1998) has described as "perspectival." This stance assumes a fundamental similarity among selves—that all kinds of selves are Is. But it also allows for a way to account for the unique qualities that characterize different kinds of beings. It involves two interlocking assumptions. First, all sentient beings, be they spirit, animal, or human, see themselves as persons. That is, their subjective worldview is identical to the way the Runa see themselves. Second, although all beings see themselves as persons, the ways in which they are seen by other beings depends on the kinds of beings observing and being observed. For example, people in Ávila say that what we perceive as the stench of rotting carrion a vulture experiences as the sweet-smelling vapor emanating from a steaming pot of manioc tubers. Vultures, because of their species-specific habits and dispositions, inhabit a different world from that of the Runa. Yet because their subjective point of view is that of persons, they see this *different* world in the same way the Runa see their own world (Viveiros de Castro 1998: 478).[19]

A tendency to see things perspectivally permeates daily life in Ávila.[20] For instance, a myth that explains why the Amazon bamboo rat[21] has such a loud call relates how this creature once asked a fallen log what women's genitals look like from its vantage point. Since such logs constitute the preferred causeways that women use to traverse their gardens, the rat figured that the log was in a privileged position to know this.[22] Alluding to the rat's abundant whiskers, the log responded, "Like your mouth." Hearing this, the rat responded, "Oh stop,"[23] and then exploded in the bawdy laughter that is now associated with its distinctive loud, long, and seemingly uncontrollable staccato call as well as its onomatopoeic name *gunguta*.[24] The humor in this myth for people in Ávila is as much about the sexually explicit reference as it is about the perspectival logic.

Another common form of perspectival joking in Ávila, as well as in other Runa communities, occurs when two people share the same name. Because I share my first name with a man in Ávila the running joke was that his wife was married to me.[25] His older sister jokingly addressed me as *turi* (sister's brother), and I addressed her as *pani* (brother's sister). Similarly, a woman who shares my sister's middle name called me brother, and one with my mother's name called me son. In all these cases shared names allowed us to inhabit a shared perspective. It allowed us to create an affectionate relationship despite the fact that our worlds are so different.

Perspectivism is certainly a historically contingent aesthetic orientation— an orientation that, pace Viveiros de Castro, we might, in this sense, describe as "cultural"—but it is also an ecologically contingent amplificatory effect of the need to understand semiotic selves in a way that simultaneously recognizes their continuity with us as well as their differences. It is a response to the challenges of getting by in an ecology of selves whose relational webs extend well beyond the human, and it emerges from everyday interactions with forest beings.

People in Ávila try to make sense of these various selves that inhabit the forest by trying to see how they see, and by imagining how different perspectives interact. One man took delight in explaining to me how the giant anteater adopts the perspective of ants in order to fool them; when the anteater sticks its tongue into ant nests, the ants see it as a branch and, unsuspecting, climb on. In their interactions with animals, the Runa, in many ways, try to emulate the anteater. They attempt to capture the perspective of another organism as part of a larger whole. This is what is involved in making a scarecrow. It is also employed in certain techniques used to catch fish. Ventura's father used to paint his hands a dark purple with the crushed fruits of *shangu*, a distant ginger relative,[26] so that armored catfish[27] would not notice his attempts to grab them from underneath the rocks and boulders in the river.

Such ecological challenges of understanding how the anteater eats ants, or how to make a scarecrow that will scare parakeets, or how to fish for catfish without being recognized by them requires an attentiveness to the points of view of other organisms. This attentiveness grows out of the fact that ants, parakeets, armored catfish, and indeed all the other life-forms that make up the rain forest, are selves. Who and what they are is, through and through, the product of the ways they represent and interpret the world around them and the ways in which others in that world represent them. They are selves, in short, that have a point of view. This is what makes them animate, and this animation enchants the world.

THE FEELING OF THINKING

People in Ávila take great pleasure in finding a viewpoint that encompasses multiple perspectives. One Ávila myth exquisitely captures this aspect of a perspectival aesthetic. It begins with a hero on top of his roof patching it. When a man-eating jaguar approaches, the hero calls out to him, "Son-in-law, help me find holes in the thatch by poking a stick through them." From the vantage point of someone inside a house it is quite easy to spot leaks in the thatch because of the sunlight that shines through them. However, because roofs are so high, it is impossible, from this position, to patch these. A person on the roof, on the other hand, can easily patch the holes but cannot see them. For this reason, when a man is patching his roof he will ask someone inside to poke a stick through the holes. This has the effect of aligning inside and outside perspectives in a special way; what can only be seen from the inside suddenly becomes visible to the person on the outside who, seeing these two perspectives as part of something greater, can now do something. Because the hero addresses and "sees" the jaguar as son-in-law, the jaguar thus hailed feels obligated to fulfill the functions incumbent on this role. Once the jaguar is inside, the hero slams the door shut and the structure suddenly turns into a stone cage that traps him.

A perspectival stance is certainly a practical tool, like the stick used to link inside and outside views, but it also affords something else. It allows one to linger in that space where, like a shaman, one can be simultaneously aware of both viewpoints as well as how they are connected by something greater that, like a trap springing shut, suddenly encompasses them. The attention people in Ávila give to such moments of awareness is a signature of Amazonian multinatural perspectivalism. This is lost when multinatural perspectivalism is taken up as a more generic analytic shorn of its shamanistic component (see, e.g., Latour 2004).

I propose that this perspectival mythic episode, in which the hero comes to unite these divergent perspectives through a vantage that encompasses them, captures, savors, and makes available something about life "itself." It captures something about the logic of the thoughts of the forest. And it captures the feeling of being alive to this living logic in moments of its emergence. It captures, in short, what it feels like to think.

Regarding this experience of coming to see inside and outside perspectives by virtue of something greater that encompasses them, consider Peirce's discussion of the experience of learning to move one's hands simultaneously and in opposite directions such that they trace parallel circular paths in the air: "To learn to do this, it is necessary to attend, first, to the different actions in different parts of the motion, when suddenly a general conception of the action springs up and it becomes perfectly easy" (Peirce 1992c: 328).

Like Peirce's example, the jaguar-trapping myth captures what it feels like when a self "suddenly" comes to see different perspectives as contributing to the more general whole that unites them. As such it calls to mind what Bateson (2002) calls "double description," which he considers central to life and mind. In thinking about double description I draw on—but simplify—Hui, Cashman, and Deacon's (2008) analysis of the concept. Bateson illustrates what he means by double description through binocular vision. By recognizing the similarities and systematically comparing the differences between what each eye sees, the brain, performing a "double description," comes to interpret each of these inputs as part of something more encompassing at a higher logical level. Something novel emerges: the perception of depth (Bateson 2002: 64–65).

Bateson asks, "What pattern connects the crab to the lobster and the orchid to the primrose and all the four of them to me? And me to you? And all the six of us to the amoeba in one direction and to the back-ward schizophrenic in another?" (2002: 7). His answer: double description is operative in the form-generating dynamics that make these entities what they are and how they are connected. The production of a series of roughly similar legs in a "proto-crab" enabled, over evolutionary time, the adaptive differentiation among these legs (some developing into claws, etc.), which allowed the organism as a whole to better "fit" or represent its environment. Just as depth emerges when the brain compares the differential duplication of ocular perspective, a crab as an organism with an overall form that fits a given niche (enabling it, for example, to walk sideways on the ocean floor) emerges over evolutionary time as an embodied interpretation of the duplication of gradually differing legs. Both involve double descriptions.

The lobster also emerges as a form that is the embodied product of a double description involving the differential duplication of appendages. Via different genetic mechanisms, the distinctive overall shape of the orchid and the primrose flower (each adapted to its respective pollinators) also results, in each case, from a double description involving the differential duplication of petals. When we compare crabs and lobsters, and these to the pair of plants, as Bateson does, we also perform double descriptions; we recognize the similarities and systematically compare the differences among these to reveal the double description that is operative in making each kind of organism what it is. When we then compare the ways we use double description to arrive at this realization with the way double description operates in the emergence of these biological forms, we see that our form of thinking is of and like the biological world; what is more, double description itself emerges as a conceptual object thanks to this higher-order double description.

Developing double description from the double description manifest in the world so that double description as a generative modality of mind becomes apparent gives us, then, the added experience of what it is like to think with the double description that is operative in the world. Or, to put it in the terms of this book: thinking with forests allows us to see how we think like forests in ways that reveal some of the sylvan properties of the living thought itself as well as how we experience these properties.

A shamanistic perspectival aesthetic cultivates and reflects on this process. In the jaguar-trapping myth a higher-order vantage "suddenly ... springs up," which connects inside and outside perspectives as elements of something greater. This allows the listener to experience the feeling of a new living thought as it emerges; it captures what it feels like to think. In Ávila this is personified in the figure of the shaman, which is the Amazonian quintessence of a self, for all selves, as selves, are considered shamans (see Viveiros de Castro 1998) and all selves think like forests.

THE LIVING THOUGHT

Lives and thoughts are not distinct kinds of things. How thoughts grow by association with other thoughts is not categorically different from how selves relate to one another. Selves are signs. Lives are thoughts. Semiosis is alive. And the world is thereby animate. People, like the Ávila Runa, who enter into and try to harness elements of a complex web of living thoughts are inundated by the logic of living thoughts such that their thoughts about life also come to instantiate some of the unique qualities of living thoughts. They come to think with the forest's thoughts, and, at times, they even experience themselves thinking with the forest's thoughts in ways that reveal some of the sylvan properties of thought itself.

To recognize living thoughts, and the ecology of selves to which they give rise, underscores that there is something unique to life: life thinks; stones don't. The goal here is not to name some essential vital force, or to create a new dualism to replace those old ones that severed humans from the rest of life and the world. The goal, rather, is to understand some of the special properties of lives and thoughts, which are obscured when we theorize humans and nonhumans, and their interactions, in terms of materiality or in terms of our assumptions (often hidden) about symbolically based linguistic relationality.

For Bateson, what makes life unique is that it is characterized by the ways in which "a difference" can "make a difference" (2000a: 459). Differences in soil can, thanks to layers of living representational relationships, come to make a difference for plants immersed in a complex semiotic ecology. And these differences can make a difference for other life-forms as well. Semiosis clearly involves differences; thoughts and lives grow by capturing differences in the world. And getting certain differences right—dogs need to be able to differentiate between mountain lions and deer—is vital.

But difference, for the living thought, is not everything. A tick doesn't notice the differences between a mountain lion and a deer, and this confusion is productive. Attending to the ways other kinds of selves inhabit and animate the world encourages us to rethink our ideas of relationality built on difference. The way selves relate is not necessarily akin to the ways in which words relate to each other in that system we call language. Relating is based neither on intrinsic difference nor on intrinsic similarity. I have explored here a process prior to what we usually recognize as difference or similarity, which depends on a form of confusion. Understanding the role that confusion (or forgetting, or indifference) plays in the living thought can help us develop an anthropology beyond the human that can attend to those many dynamics central to living and thinking that are not built from quanta of difference.

THREE

Soul Blindness

Out of sleeping a waking,
Out of waking a sleep;
Life death overtaking;
Deep underneath deep?

—Ralph Waldo Emerson, *The Sphinx*

Ramun, the schoolteacher's ten-year-old brother in-law, pitched his skinny mass out of Hilario's doorframe and called out earnestly, "Pucaña!" By now we were pretty sure that something had gone wrong. Pucaña and Cuqui still hadn't come home. We didn't yet know that they had been killed by a feline, but that was what we were starting to suspect. Huiqui had straggled in moments earlier with a gaping hole at the back of her head. Hilario was patiently cleaning her wound with some rubbing alcohol from my first-aid kit. Ramun still harbored some hope that Pucaña would turn up. And so he called out her name once more. When she didn't appear he turned to us and said, "What's-its-name. I'm calling the one that's become shit." América responded, "She must have become shit. That's what jaguars do. They just shit them out."[1]

After retracing our steps to the patchwork of forest and fallows where the women had been harvesting fish poison and where they had heard the dogs' last barks, we finally found their bodies. The dogs had indeed been killed, if not exactly eaten, by a feline, which the family would later conclude was a jaguar and not the mountain lion that the women had originally imagined the dogs had mistaken for a deer. Huiqui would not make it through the night.

Selves, like Pucaña, or like us, are ephemeral creatures. They can come to inhabit ambiguous spaces—no longer fully interactive subjects that can be named and that, like Pucaña, can also potentially respond to these names, or quite yet transformed into inanimate objects like dead meat, *aicha*, or jaguar shit. Nor, for that matter, can they fully inhabit that final space of silence; *chun* is the word Luisa used to describe it. Rather, selves can come to be caught somewhere in the space between life and death, somewhere in that ambiguous space of "what's-its-name" (*mashti*, in Quichua),[2] of the almost nameless— not exactly here with us but not fully elsewhere either.

This chapter is about the kinds of spaces and transformations, the flip-flops, the difficulties, and the paradoxes captured by the word *mashti*. It is about the different ways in which selfhood can dissolve and the challenges this poses for beings living in an ecology of selves. Such dissolutions come in many forms. There is, of course, the catastrophe of organismic death. But there are also many kinds of disembodiments, and many ways in which selves can become reduced from a whole to an objectlike part of another self. And, finally, there are ways in which selves can break down as they lose the ability to perceive and interact with other selves as selves.

This chapter is also about selves and objects and their co-constitution, and it is especially about how selves create objects and how they can also become objects. And it is about the difficulties this fact of life poses for us, as well as what an anthropology beyond the human can learn about such difficulties, thanks to the peculiar ways in which such difficulties become amplified in this particular ecology of selves of the Ávila region.

Although the beginning of life on this earth surely represents, as Jesper Hoffmeyer (1996: viii) so nicely phrased it, the moment when "something" became "someone," that something did not exactly exist before there was a "someone." It is not so much that things didn't exist before there were beings to perceive them but rather that before living thoughts emerged on this earth nothing ever came to stand in relationship to a self as an object or as another. Objects, like selves, are also effects of semiosis. And they emerge out of semiotic dynamics that exceed the human.

This chapter, then, is about the various dissolutions of self that living creates. It is about what Stanley Cavell (2005: 128) calls the "little deaths" of "everyday life"—the many deaths that pull us out of relation. That death is such a central part of life exemplifies what Cora Diamond (2008) calls a "difficulty of reality." It is a fundamental contradiction that at times overwhelms us humans with its sheer incomprehensibility. And this is compounded by another difficulty: such contradictions are at times, and for some, completely

FIGURE 6. When dead animals are brought home from the hunt they are fondled with curiosity by children and studiously ignored by adults. Photo by author.

unremarkable. The feeling of disjunction that this lack of recognition creates is also part of the difficulty of reality. Hunting, in this vast ecology of selves, in which one must stand as a self in relation to so many other kinds of selves who one then tries to kill, brings such difficulties to the fore; the entire cosmos comes to reverberate with the contradictions intrinsic to life (figure 6).

LIFE BEYOND THE SKIN

The particular configuration of matter and meaning that constitutes a self has a fleeting existence. Pucaña and the other dogs in some real sense ceased being selves the moment they were killed by the jaguar. Living selfhood is localized around such fragile bodies. To say that a self is localized, however, does not mean that it is necessarily or exclusively inside a body, "shut up in a box of flesh and blood," as Peirce critically put it (CP 7.59; see also CP 4.551), or "bounded by the skin," in Bateson's words (2000a: 467). Life also extends beyond the confines of one particular embodied locus of selfhood. It can potentially exist in some sort of semiotic lineage thanks to how selves are represented by other selves in ways that matter to these subsequent selves.

Beyond individual death there is, then, a kind of life. And the generality of life, its potential to spread into the future, in fact, depends on the spaces that such singular deaths open up (see Silverman 2009: 4). Ventura's mother, Rosa, died while I was living in Ávila. But she did not altogether cease being. According to her son, she entered "inside" (ucuman) the world of the spirit masters— the beings who own and protect the animals of the forest (see chapters 4–6)— and she married one of them. All that was left of her in the "above" world (jahuapi), the world of our everyday experience, was her "skin." According to Ventura, his mother "just discarded her skin"[3] when she went to the spirit world, and this skin was what was left for her children to bury at her funeral. Rosa lived on, outside her old skin, forever, as a timeless nubile bride in the world of the masters.

We will all eventually cease being selves. And yet traces of that unique configuration that constitutes what we take to be our selfhood can potentially exceed our mortal skin-bound bodies and in this manner "we" might persist, in some form, well after the end of our "skins." As I argued in chapter 2, selves are outcomes of semiosis. They are embodied loci of interpretant formation—the process by which one sign is interpreted by another in a way that gives rise to a new sign. Selves, then, are signs that can potentially extend into the future insofar as a subsequent self, with its own embodied locus, re-presents it as part of that semiotic process by which that subsequent self emerges as a self. Life, then, without ever being fully disembodied, potentially exceeds any skinbound self around which it might currently be localized. Death, as I will argue, is central to the ways a self exceeds its current embodied limits.

Selves exist simultaneously as embodied and beyond the body. They are localized, and yet they exceed the individual and even the human. One way to capture this way in which selves extend beyond bodies is to say that selves have souls. In Ávila the soul—or alma as people call it, using a term of Spanish origin—marks the ways in which semiotic selves are co-constituted in interaction with other such selves. Souls emerge relationally in interaction with other souled selves in ways that blur the boundaries we normally recognize among kinds of beings.

Having an alma is what makes relation possible in the ecology of selves that the Ávila Runa inhabit. Because, according to people in Ávila, animals are "conscious"[4] of other kinds of beings, they have souls. For example, both the dog and the agouti, a large, edible forest rodent that, along with the peccary, is considered quintessential game (aicha in Quichua), possess souls because of their abilities to "become aware of,"[5] to notice, those beings that stand in relation to them as predator or prey. The agouti is able to detect the presence of its predator the dog, and therefore it has a soul. This relational capacity is reified; it has a physical location in the body. The agouti's gall bladder and sternum serve as its organs of consciousness. Through these, the agouti detects the presence of predators. People's awareness of other beings is also somatically localized. Muscular twitches, for instance, alert them to the presence of visitors or dangerous animals such as poisonous snakes.

Because the soul, as relational quality, is located in specific parts of the body, it can pass to others when these parts are eaten. Dogs are defined as conscious, soul-possessing beings because of their ability to detect agoutis and other game. They can increase their consciousness—as measured by their increased ability to detect prey—by ingesting the very organs that permit the agouti to detect the presence of dogs. For this reason people in Ávila sometimes feed the agouti's bile or sternum to their dogs.

Following the same logic, they also increase their consciousness of other beings by ingesting animal body parts. Because bezoar stones, the indigestible accretions sometimes found in deer stomachs, are considered the source of deer's awareness of predators, hunters sometimes smoke their scrapings in order to encounter deer more readily. Some people in Ávila become runa puma by drinking jaguar bile; this helps them adopt a predatory point of view, and it facilitates the passage of their souls into the bodies of jaguars when they die.

Like people in Ávila, Peirce saw the soul as a marker of communication and communion among selves. He saw the soul as capturing certain general properties inherent to a living semiotic self in constitutive interaction with other such selves.[6] Accordingly, Peirce locates the "seat of the soul," not necessarily in a body, even though it is always related to a body, but as an effect of intersubjective semiotic interpretance: "When I communicate my thought and my sentiments to a friend with whom I am in full sympathy, so that my feelings pass into him and I am conscious of what he feels, do I not live in his brain as well as in my own—most literally?" (CP 7.591). The soul, according to Peirce, is not a thing, with a unitary localized existence, but something more like a word, in that its multiple instantiations can exist simultaneously in different places.

Living thoughts extend beyond bodies. But this fact poses its own problems. Just how do selves extend beyond the limits of the bodies that house them? And where and when do such selves finally come to an end? How life extends beyond bodies in such a way that somehow entangles selfhood with the fact of finitude is a general problem. It is a problem inherent to life, and it is one that this ecology of selves amplifies in ways that might allow an anthropology beyond the human to learn something about the way that death is intrinsic to life.

In Ávila this problem becomes particularly salient in the interactions people have with runa puma. Were-jaguars are ambiguous creatures. On the one hand, they are others—beasts, demons, animals, or enemies—but, on the other, they are persons who retain powerful emotional connections and a sense of obligation to their living relatives.

This ambiguous position poses serious challenges. Ventura's recently deceased father's puma killed one of his son's chickens. This angered Ventura and made him doubt whether his father, now a jaguar, still continued to consider him a son. Accordingly, Ventura went out to the woods near his house and spoke out loud to his father, who was around there, somewhere, inhabiting the body, and the viewpoint, of a jaguar:

> "I'm not an *other*," I told him.
> "I'm your son."
> "Even when I'm away,
> you need to look after my chickens."[7]

He continued to criticize his father for not acting more like a real puma who, instead of snatching chickens, should be out in the deep forests hunting for himself: "'Is that what you're gonna do instead of going off to the mountains?'" "'If you're gonna stick around here,'" Ventura continued, "'you need . . . to catch at least something for me.'" Shortly after—"It wasn't long— I think it was only about three days"—Ventura's father's puma finally began to fulfill his obligations: "Just like that, he gave me a nice agouti he caught."

This is how Ventura came upon the "gift" from his father. He first discovered the kill site in some brush near his house. He observed that the jaguar had "trampled" a clearing "until it was shiny." From this shiny clearing Ventura followed the trail made by the jaguar pulling the carcass through the brush.

> And then I saw
> this,
> this here head, just a head cut off.
>
> . . .
>
> After that, I looked around and noticed a string of entrails

. . .

And then the puma dragged it even further

Ventura, gesturing with his hands, described the quarry he finally came upon.

The whole thing, from here on up was eaten.
But both legs were still good.

Not only did his father's puma leave the prime cuts for his son, but he also wrapped them, just like the gifts of smoked meat presented to invited kin at a wedding.

Covering it with leaves.
Wrapping it up inside them,
he just left it.

The puma's gift is a half-eaten, disemboweled agouti carcass—a body no longer recognizable as a self but now transformed into cuts of packaged meat.

Were-jaguars are ambiguous creatures. One is never sure if they really are still human. Will they forget to fulfill the duties of a relation? And when they are encountered in the forest in all their ferocious otherness might they not also simultaneously be the kind of person to whom *we* owe obligations?

One day out hunting, Juanicu happened upon a jaguar. He shot at it with his small muzzle-loading shotgun, a gun that is not very effective against large felines. This is how, with nothing more than a cascading chain of iconic sound images, he re-created the event:

tya
(a gun firing successfully)

tsi'o—
(the vocalization made by the jaguar as it was hit)

tey'e—
(the ammunition hitting its target)

hou'u—[b]
(another vocalization made by the jaguar)

Then, rapidly and somewhat more softly, Juanicu imitated the sound made by the lead shot hitting the jaguar's teeth:

tey tey tey tey

The shot shattered the jaguar's teeth and severed some of his whiskers. After the jaguar ran off, Juanicu picked up some of the whiskers that had been blasted off, shoved them—"huo'"—into his pocket, packed up the jaguar's half-eaten quarry, and went home.

That evening, the jaguar was still with him. "He made me dream," Juanicu told me, "all night long." In those dreams Juanicu's long-dead *compadre* came to him and appeared just like he had when he was still alive, except that when he opened his mouth to talk his shattered teeth were visible: "'How is it that you can do such a thing to a *compadre*?'" he asked Juanicu. "'Now what am I gonna eat with?'" Juanicu's compadre then paused and panted, "'hʰa–,'" the way jaguar's do, and then he continued, "'Like this, I won't be able to eat. Like this I'm gonna die.'" "And that," Juanicu concluded, "is how he told me what happened . . . that's how the soul tells you at night when you dream." After a long pause, Juanicu added, "I shot *that*. I sent *that* off."[8]

The runa puma is a strange creature; he reveals himself as a compadre and yet pants like a jaguar. Juanicu is bound to *him* through ritual kin ties, and yet he has no remorse about shooting *it*. The runa puma who spoke to Juanicu is a self; the selfsame one he shot is a thing.[9]

This contradictory nature of the puma also came up in the conversations Hilario and his family had about the identity of the jaguar that killed their dogs. Several hours after Ramun called out to Pucaña, the family found her body out in the forest strewn beside Cuqui's and concluded, from the tracks in the area and the bite marks to the backs of their heads, that it had been a jaguar that had killed them.

But they still didn't know what *kind* of jaguar was responsible. They suspected it was a runa puma and not just a regular "forest jaguar" (*sacha puma*), but this, in and of itself, was not a fully satisfying answer. As one family member put it, "Whose puma would bother us like this?" That night they got their response. Everyone dreamed of Hilario's dead father. Amériga dreamed that her father-in-law came up to her wearing a hat and asked her to store a large package of game meat he had been given. Luisa dreamed that she could see her father's testicles and that his intestines were coming out of his anus. Later that evening she dreamed of two calves, one black and one mottled, which, she reasoned, must belong to her father, now himself a master in the afterlife realm of the spirit masters of the forest (see chapter 6).

Hilario's son Lucio was away from home. He had not heard news of the attack from his family and didn't return until the day after it happened. But he too had dreamed that night of his grandfather, "right there just talking and laughing with me." This, for him, secured the jaguar's identity: "So it must have been my dead grandpa—so it must have been him wandering around." It must have been, that is, his grandfather's soul, in a jaguar's body, wandering the thickets near the house, seeing the world through jaguar eyes, seeing the family's dogs as prey.

Lucio didn't dream of a fierce jaguar but of a loving grandfather. He and his grandfather were together, talking and laughing.[10] Laughter, like crying and yawning, is contagious. It provokes laughter in others and, in this way, unites them, through a kind of iconism, as one in a shared sentiment (see Deacon 1997: 428–29). It unites them, in Peirce's words, in a "continuity of reaction" (CP 3.613). As they laughed together Lucio and his grandfather, for a moment, formed a single self in communicative communion.

But as far as Hilario and his family could tell this jaguar—the beloved grandfather—attacked the dogs for no good reason. Some runa puma attack dogs when their relatives don't observe the taboos that are prescribed after the death of a relative. This was not the case here. And this made the attack incomprehensible. For Lucio, this were-jaguar was "no good." For Hilario, he was "a demon," a "supai." "What else," he asked, "could it be?" "Yeah," Luisa elaborated, "transformed into a demon." América, always questioning, always wanting to know why, asked no one in particular, "How is it that, being a person, he could turn into such a creature?" Souls, as América intimated, are persons, like us, and they interact that way with us in dreams. Yet as jaguars in the forest, they might become an other kind of being—a kind of being no longer capable of sharing or caring, a kind of being that is less than dead, one that is soulless, a nonperson.

Lucio's dream-time contact with his beloved grandfather and the presence of that demonic jaguar in the forest are one and the same. "The reason I dreamed like that," Lucio reflected, "was that he must have come down for a visit." América agreed. Were-jaguars are supposed to be up in the mountains, far from where people live. It was because Lucio's grandfather had come down from his forest abode that his soul and that of his grandson could come together in laughter the night that Lucio dreamed. This also, in a way, explained the attack on the dogs.

Later that evening at his parents' house, Lucio recalled a recent encounter in the forest with a jaguar, and given the circumstances and his dream he came to the conclusion that this too was a manifestation of his grandfather. Lucio wanted to kill this puma. In his recollection he makes it "killable" (Haraway 2008: 80) by describing it as a thing, not a person. He used the inanimate pronoun *chai* (that), in its abbreviated form *chi*, instead of the animate *pai*, which in Quichua would be used to mark the third person regardless of gender or status as human:

> *chillatami carca*
> that's the one!

And he was angry that his gun malfunctioned and he missed a shot: "Damn!"

Lucio didn't regret having tried to kill this jaguar, even after learning that it harbored his grandfather's soul. His grandfather, who, in Lucio's dream, was more than a third person—was in fact a kind of *we*, united with Lucio in laughter—became for him a mere thing.

FINALIZING DEATH

The boundaries between life and death are never perfectly clear. There are moments, however, when they need to be made so. When a person dies, his or her soul—or souls, for these, like Peirce's, can be multiple and can exist simultaneously in different places—leaves the body. As with Lucio's grandfather's soul, it can enter the body of a jaguar, or it can "climb up" (*sican*) to the Christian heaven, or it can become a master in the realm of the spirit masters of the animals.

What is left is the *aya*. *Aya* in Ávila Quichua means two things. In one sense it simply means the inanimate corpse, the bag of skin that Rosa left behind for Ventura and her other children to bury. In another sense it refers to the wandering ghost of the dead, bereft of both body and soul. The soul imputes consciousness and the attendant ability to resonate and empathize with other beings. The fact that the aya has no soul makes it particularly damaging to people. It becomes *"shican,"* that is, "another kind"[11] of being—one that is "no longer capable of loving people," as one person explained it to me.[12] This is especially true of the relation it has to its family. It no longer recognizes relatives as loved ones. The aya are doubly estranged from babies born after their deaths, for their relation to them is even more tenuous. These babies are therefore quite susceptible to illnesses caused by them. Although the aya lack consciousness and a soul, they wander the places they used to frequent when alive, trying hopelessly to reattach themselves to the world of the living. By doing this, they cause sickness to their family through a kind of *"mal aire"* known as *huairasca*.

The aya inhabit a confused space. We know they are dead, but they think they are still alive. Accordingly, two to three weeks after a person dies and is buried, a ritual feast, known as *aya pichca*,[13] is held in order to rid the living of the dangers of the aya still in their presence and in this way to definitively separate the realm of living selves from that of the lifeless. This ritual begins in the early evening and lasts well into the next morning. This is followed by a special meal (see chapter 4). Such an aya pichca was held after Jorge, Rosa's husband and the father of Ventura, Angelicia, and Camilo, died. The first part began in the early evening and lasted the entire night, until just before dawn. It consisted of a drinking party in Jorge's abandoned house.

Although there was some crying and some of the distinctive chantlike wailing that often accompanies mourning in Ávila, the mood for the most part was joyous. In fact, Jorge was treated as if he were still alive. When Jorge's daughter Angelicia arrived at his house, she left beside the bed he once slept on a bottle of the home brew *vinillu*, saying, "Here, drink this sweet water."[14] Others would later serve him bowls of fish soup. When a neighbor placed a bottle of vinillu on the bench, another fell off. This prompted someone to remark that Jorge, now a little drunk, was knocking over bottles. As we were about to go to Camilo's house nearby, Angelicia's husband, Sebastián, said, "OK, Grandpa, you just wait, we'll be back in a bit."[15]

Despite the ways in which people treated Jorge as if he were still part of an intimate social circle of the living—joking with him, talking to him, sharing food and drink with him, taking temporary leave and then returning to immerse him in a final all-night party—the purpose of this ritual was actually to send Jorge's aya off, definitively and forever, to reunite with his afterbirth (*pupu*) buried back near the Huataracu River, where his parents lived at the time of his birth.[16] Only when that empty remnant of self, marked by the aya, is realigned with the placental trace marking Jorge's emergence as a unique embodied locus of self, will his ghost cease its dangerous wanderings.

We stayed up all night, drinking and joking beside Jorge's bed. As daylight approached, a time when Jorge would have normally gone off hunting, the mood changed. Someone came around and painted our faces with achiote. A dab of this reddish orange face paint served as a kind of cloak that made our nature as human selves invisible to Jorge's aya. No longer able to see us as persons, he would be unaware of our presence, and, in this manner, he would not be detoured from his resting place.

This is how it must be. The aya are extremely dangerous to the living, and unmediated intersubjective encounters with them, such as seeing or speaking with them, can cause death. For such encounters require seeing the world from the point of view of these nonliving, nonselves. And this, in turn, would imply the radical dissolution of our selfhood—something we would not be able to survive.

Our faces now painted with achiote, we took basketfuls of Jorge's possessions outside and placed them on a path that Jorge's aya would walk to reunite with his afterbirth. Children were notably present, and they were encouraged to talk to Jorge as if he were alive, urging him to go on his way with phrases like "Come on, let's go." Meanwhile, Jorge's close relatives got off the trail and hid in the forest. In this manner the aya, now unable to recognize his family, friends, and neighbors, was fanned along on its way with the leaves of *aya chini*, a giant anomalously nonstinging variety of nettle.[17] Some felt a breeze as Jorge's aya departed. His hens, placed in one of his carrying baskets, became frightened, indicating the presence of the departing aya.

At the beginning of the evening Jorge, although dead, was still a person to his living relations, someone with whom his relatives that night ate and drank and laughed and talked. By the end of the evening, however, Jorge had become excluded from that realm of commensality. He was sent forever to the separate social and relational domain of the deceased.

DISTRIBUTED SELFHOOD

Desubjectivization is not only caused by the physical dissolution of the embodied locus of selfhood in death. There are also important ways in which selves that are still living cease being treated as selves by other selves. Although people in Ávila recognize dogs as selves in their own right, they also, on occasion, treat them as tools. They sometimes compare dogs to guns, the implication being that like these "arms" dogs are extensions of human hunting abilities. People in Ávila are careful to observe special precautions regarding the implements that help them hunt. For example, they make sure that any bones from animals they have killed are disposed of in the nearby washing and drinking streams, lest the gun or trap used to kill these animals become "ruined" (*huaglirisca*).

Dogs are also subject to such potential defilement. Hilario's family was careful not to feed the dogs the large bones of the deer they had killed that week before they were attacked. The bones were instead properly discarded in the stream. In this case, because the dogs—rather than a gun or trap—had killed the deer, they might also become "ruined." Their noses, Hilario remarked, "would become stopped up,"[18] and they would no longer be able to be aware of the game animals in the forest. Dogs, then, in certain contexts are like guns. They become extensions—arms—that expand the locus of human selfhood.

People can also become thinglike tools. They can become parts of a greater whole, appendages of a larger self. At a drinking party, Narcisa, in her early twenties, told us of an encounter she had had the day before with a doe, a buck, and their fawn in the woods near her house. Deer are coveted game animals, and Narcisa was hoping to kill one. But there were a couple of problems. First, women don't usually carry guns, and she regretted that she was unarmed. "Damn!" she exclaimed, "If I had that thing"—that is, a shotgun— "it would've been great!"[19] Second, her husband, who did have his gun handy and was in the vicinity, hadn't seen the deer. Fortunately, however, the night before Narcisa had, as she put it, "dreamed well." And this led her to think that they would be able to get one of those deer.

Narcisa was faced with the challenge of trying to alert her husband to the presence of deer without at the same time alerting the deer to her own presence. She attempted to "yell" forcefully but at the same time quietly by substituting an increase in volume with an increase in word elongation:

"'Aleja—ndru,' I quietly cried out."

The tension in her throat absorbed the volume of the sound without decreasing the urgency of her message. She was hoping, in this way to remain inaudible to the deer. But her attempt failed:

after calling like that
the doe noticed
and slo—wly, turned around [about to run off]

More accurately, Narcisa's attempt to keep the deer from noticing her only partially failed. The buck, as opposed to the doe, "never noticed anything."

Narcisa's challenge concerning how to selectively communicate to her husband about the deer without the deer noticing points to the ways in which agency becomes distributed over different selves and how some of these selves can lose agency in the process. Narcisa is the primary agent here. Dreaming is a privileged form of experience and knowledge, and it was she, not her husband, who had dreamed. Narcisa's "good dreaming" was the important action. Her husband's ability to shoot the animal was simply a proximate extension of this.

Narcisa's agency is the locus of cause—it is her dream that counted—and yet her intentions can only be successfully realized by extending herself through objects. Without a gun, she can't shoot a deer, and because men generally are the ones who carry guns in Ávila, she must involve her husband. In this context, however, he is not really a person but rather, like a gun, he becomes an object, a tool, a part through which Narcisa can extend herself.

The distribution of selves and objects in this situation should, Narcisa hoped, have looked as follows: Narcisa and Alejandro should have been united as a single individual in a "continuity of reaction," oriented, together, as predator toward the killing of a deer, here thought of as a prey object. Narcisa and Alejandro, in other words, should have become an emergent single self, whereby two selves become one by virtue of their shared reaction to the world around them (see Peirce CP 3.613). For such a "continuity of being" (CP 7.572), as Peirce has it, creates "a sort of loosely compacted person, in some respects of higher rank than the person of an individual organism" (CP 5.421). This emergent self need not have been equally distributed. Narcisa would have been the locus of this agency, and Alejandro, like Hilario's dog, would have become an arm—an object through which Narcisa extended her agency.

But things did not turn out this way. The continuity of reaction oriented itself, not along species lines, but along gender ones, and these crossed species boundaries in ways that disturbed the particular predator/prey distribution that Narcisa had hoped for. The doe noticed Narcisa. Neither the buck nor the husband ever noticed anything. This is not the way Narcisa wanted things to turn out. Narcisa and the doe here were the sentient selves, united, inconveniently, it turned out, through a continuity of being as a higher-order single self. In "never noticing anything," the males had become objects.

SEEING BEYOND ONESELF

Alejandro and the buck remained unaware of those other selves in their presence. This is dangerous. If trans-species interactions depend on the capacity to recognize the selfhood of other beings, losing this capacity can be disastrous

for beings, such as these two males, who are caught up in the webs of predation that structure this forest ecology of selves. Under certain circumstances
we are all forced to recognize the other kinds of minds, persons, or selves that
inhabit the cosmos. In this particular ecology of selves that entangles Alejandro and the buck, selves must recognize the soul-stuff of other selves in order
to interact with them.

That is, in this ecology of selves, to remain selves, all selves must recognize
the soul-stuff of the other souled selves that inhabit the cosmos. I've chosen
the term *soul blindness* to describe the various debilitating forms of soul loss
that result in an inability to be aware of and relate to other soul-possessing
selves in this ecology of selves. I adopt the term from Cavell (2008: 93), who
uses it to imagine situations in which one might fail to see others as humans.[20]
Because in this ecology of selves all selves have souls, soul blindness is not just
a human problem; it is a cosmic one.

Soul blindness, in this Ávila ecology of selves, is marked by an isolating
state of monadic solipsism—an inability to see beyond oneself or one's kind. It
arises when beings of any sort lose the ability to recognize the selfhood—the
soul-stuff—of those other beings that inhabit the cosmos and it emerges in a
number of domains. I enumerate a few examples here to give a sense of the
range and prevalence of this phenomenon. For instance, something known as
the hunting soul[21] allows hunters to be aware of prey in the forest. Shamans
can steal this soul with the effect that the victim can no longer detect animals.
Without this soul, hunters become "soul blind." They lose their ability to treat
prey-beings as selves and can therefore no longer differentiate animals from
the environments in which they live.

Hunting is also made easier by the soul loss of prey. Men who kill the souls
of animals in their dreams can easily hunt them the next day because these
animals, now soulless, have become soul blind. They are no longer able to
detect their human predators.

Shamans do not only potentially steal the souls of hunters, they can also
steal the souls of the vision-producing aya huasca plants of their shamanic
rivals with the effect that these plants become soul blind; ingesting them no
longer permits privileged awareness of the actions of other souls.

The invisible darts through which the shaman attacks his victims are propelled by his soul-containing life breath (*samai*). When darts lose this breath
they become soul blind; they are no longer directed at a specific self but travel
aimlessly, without intention, causing harm to anyone that happens upon their

path. Jorge's aya was soul blind in a manner very similar to the shaman's spent darts, it lacked the ability to engage in normative social relationships with its living relatives, and was therefore seen as dangerous.

Adults sometimes punish children by pulling at tufts of their hair until a snapping sound is made. These children become temporarily soul blind; they become dazed and unable to interact with others.

The crown of the head, especially the fontanel,[22] is an important portal for the passage of life breath and soul-stuff. Soul blindness can also be effected by extracting life breath through the fontanel. Delia described the jaguar that killed the dogs as having "bit them with a *ta'* on their animal-following crowns."[23] *Ta'* is an iconic adverb, a sound image, that describes "the moment of contact between two surfaces, one of which, typically, is manipulated by a force higher in agency than the other" (Nuckolls 1996: 178). This precisely captures the way in which the jaguar's canines impacted and then penetrated the dogs' skulls. That people in Ávila consider such a bite lethal has much to do with the ways in which this part of the body permits intersubjectivity. The dogs' deaths, then, were the result of a complete loss of their "animal-following" capabilities—the radical and instantaneous imposition of soul blindness.

Some notion of the motivations of others is necessary for people to get by in a world inhabited by volitional beings. Our lives depend on our abilities to believe in and act on the provisional guesses we make about the motivations of other selves.[24] It would be impossible for people in Ávila to hunt or to relate in any other way within this ecology of selves without treating the myriad beings that inhabit the forest as the animate creatures that they are. Losing this ability would sever the Runa from this web of relations.

PREDATION

Hunting within an ecology of selves is tricky business. On the one hand, the sharing of food and drink, and especially of meat, is, throughout Amazonia, crucial to the creation of the kinds of interpersonal relations that are the basis for community. Growing children should have plenty of meat, and their grandparents and godparents should also receive regular gifts of meat. Relatives, compadres, and neighbors who come to help clear forest and build houses also need to be fed meat. Sharing meat is central to the fruition of social ties in Ávila. And yet that meat that is shared and consumed was also, at one point, a person. Once one recognizes the personhood of animals, there is always the danger of confusing hunting with warfare and commensality with cannibalism.[25]

To notice and to relate to the various beings that live in this ecology of selves, these various beings must be recognized as persons. But to eat them as food, they must eventually become objects, dead meat. If the selves that are hunted are persons, then might not people too eventually become dehumanized objects of predation? Jaguars do, in fact, sometimes attack hunters in the forest. And sorcerers can assume the appearance of predatory raptors. This is why, as Ventura noted, one should never try to kill an agouti that runs into the house, for it is surely a relative, transformed into the fleeing prey of a predatory sorcerer that has taken the form of a raptor. Predation points to the difficulties involved when selves become objects or treat other selves as objects within an ecology of selves.

As I mentioned, at times people consume animals, not as meat, but as selves, to acquire some of their selfhood. Men drink jaguar bile to become puma, and they feed agouti sternums and other soul-containing body parts to their hunting dogs. These substances are consumed raw to preserve the selfhood of the creature being eaten. This, as Carlos Fausto (2007) has noted, amounts to a kind of cannibalism. By contrast, when people want to eat commensally, that is, when the communion is not with the eaten but among the eaters, then the eaten must be transformed into an object. Processes of desubjectivization, such as cooking, are central to this, and the Ávila Runa in this regard are like so many other Amazonians in thoroughly boiling their meat and avoiding cooking processes such as roasting that can leave some of the meat raw (Lévi-Strauss 1969).

An ecology of selves is a relational pronominal system; who counts as an *I* or a *you* and who becomes an *it* is relative and can shift.[26] Who is predator and who is prey is contextually dependent, and people in Ávila take great relish in noting how these relationships can sometimes become reversed. For example, a jaguar trying to attack a large land turtle (*yahuati*) is said to have gotten its canines caught in the turtle's carapace and was forced to abandon not only his prey but also his teeth that had broken off and remained lodged in the turtle's shell. Now toothless, the jaguar was unable to hunt and soon began to starve. When the jaguar finally expired, the turtle, that great lover of carrion, with the jaguar's canines still impaled in its shell, began to eat the rotting flesh of its former predator. The jaguar was thus transformed into its former prey's prey. This quintessential *I* is only so by virtue of the relationship it has to an *it*—to aicha, or prey. When this relationship changes, when the turtle becomes a puma, the jaguar is no longer the predator. Jaguars are not always jaguars; sometimes turtles are the real jaguars. What *kind* of being one comes to be is the product of how one sees as well as how one is seen by other kinds of beings.

Because trans-species relationality is so overwhelmingly predatory in this cosmic ecology of selves, those creatures that don't neatly fit are especially interesting. One class of beings that receives such attention is the mammalian order Xenathera, which includes such seemingly disparate creatures as sloths, anteaters, and armadillos. Another name for this order in the Linnean system is Edentata. Appropriately, this means "rendered toothless" in Latin, and it alludes to one of the most striking features that makes this group a *kind*, both for biologists and for people in Ávila: its members lack "true" teeth; they develop no milk teeth and lack canines, incisors, and premolars. Members of this order have only peglike teeth, if they have any at all (Emmons 1990: 31).

Teeth are central markers of predator status. Hilario once told us of an enormous jaguar that people in Ávila managed to kill many years ago. The canine teeth were the size of small bananas, and, according to him, the village women, imagining how many people those teeth must have killed, wept when they saw them. Because canines embody the essence of a predatory nature, people use jaguar canines to put hot pepper in the eyes of children so that they too will be pumas. Without their canines, jaguars are no longer pumas. Jaguars, people say, die when their teeth wear out.

It is in this context that the members of the "toothless" order are so salient. Legend has it that the collared anteater (*susu*) is prone to fighting with the sloth (*indillama*), saying, "You have teeth and still you have thin arms. If I had teeth I would be even fatter than I already am." Sloths have vestigial peglike teeth; the arboreal collared anteater, like its larger terrestrial cousin the giant anteater, or tamanuhua, completely lacks teeth. Despite their lack of teeth, anteaters are formidable predators. An arboreal anteater can easily kill a dog, and it is indefatigable. It is known to withstand many shots before it falls to the ground, and once on the ground a hunter will often have to pound on its head with a stick to kill it. The giant anteater is considered a puma in its own right. Though it lacks teeth, its sharp claws can be lethal. Juanicu was almost killed by one while I was living in Ávila (see chapter 6). Even the jaguar is said to be afraid of the giant anteater. According to Ventura, when a jaguar encounters a giant anteater sleeping between the buttresses of a tree he will signal for all to be quiet, saying, "'Sshh, don't tap [the buttress], big brother-in-law's sleeping.'"[27]

Because armadillos lack true teeth they also don't easily fit into the predator/prey ecological cycle of self-perpetuation through object creation. In contrast to the anteaters, armadillos are not at all aggressive, and by no means can they be construed as threatening predators. This is how Emmons (1990: 39) describes their innocuous nature: "[They] trot with a rolling or scuttling gait, some like windup toys, snuffling and grubbing with their noses and forepaws and seemingly unaware of anything more than a foot or two away."

Armadillos have their own kind of spirit master, the *armallu curaga*, or Lord of the Armadillos, who owns and protects them. Appropriately, the entry to this lord's home is a tunnel, like that of an armadillo's burrow. Legend has it that an Ávila man got lost in the forest and was eventually found by this master, who then invited him home to share a meal. When the food was brought out the man saw piles of freshly cooked, steaming-hot armadillo meat. The master, by contrast, saw this same food as cooked squash. Like a squash, the armadillo has a hard "rind." What from our vantage appears as this animal's intestines, the master sees as a tangled mass of seeds enveloped by the fibrous and sticky flesh at the heart of a squash.

Like his armadillos, the lord had no teeth and, to the man's surprise, proceeded to "eat" the food before him by simply inhaling through his nose the vapor that emanated from the cooked servings. When he was finished, the food still looked to the man like perfectly good, intact cuts of meat. But the armadillo master, having already consumed all their life force, considered these cuts excrement and, to the man's dismay, discarded them.

The spirit masters of the forest, such as the armallu curaga, are predatory, like jaguars, and they are sometimes considered demonic. However, instead of eating meat and blood as jaguars and other demons do, the Lord of the Armadillos "eats" only life breath because it lacks the teeth that are the markers of a "true" predator. Unlike the jaguar through whose body Ramun imagined Pucaña being transmuted into shit, this strange predator lacks the teeth to eat meat. Therefore he doesn't shit real shit, and that process of desubjectivization is never completed. What excrement this master does produce, he smears on himself as face paint.

The master keeps his armadillos in his garden, and, as one does with squashes, he taps on them to determine if they are "ripe" and ready to eat. The Lord of the Armadillos was kind to the lost man and invited him to take one of these "squashes" home. But every time the man tried to grab one it would scuttle off—vine, leaves, and all.

People on occasion attempt to harness the fact that such predator-prey relationships are potentially reversible. Men sometimes do so by means of charms (*pusanga*), which they employ to attract and seduce animals, and sometimes women. When men use these, they want to disguise their intentions. It is fitting, then, that the most important of these charms is made from the anaconda's skull and teeth. The anaconda, along with the jaguar, is a feared predator. But unlike the jaguar, the anaconda captures its prey by a process of attraction and seduction. It causes animals and people alike to become lost in the forest. The victims, in a sort of hypnotic state, begin to wander around in circles that spiral increasingly inward until they eventually end up at the spot where the anaconda is hiding, waiting to crush them with her embrace. The anaconda is the kind of predator that hunters would like to be: one that is not initially recognized as such.

Of the various organisms that are used as ingredients for hunting or love charms, certainly the metallic-blue–colored whiplash beetle, which Juanicu calls *candarira*,[28] is among the most visually stunning. On a collecting trip in the woods with him I once pulled back a mat of leaf litter to discover a dazzling pair of the shiny slender beetles endlessly circling one another. The pulverized remains of these insects, according to Juanicu, can be placed in the food or drink of a woman one wishes to attract. The woman who comes under the spell of this charm will madly follow the man who is responsible. The insects can also be placed in a hunting bag, to attract peccaries to the hunter. In the endless way in which they circle one another, like the serpent Oroborus biting its tail, these insects bind predator and prey into one, such that their roles become confused. This is seduction; the prey is now predator, and the original predator incorporates this apparent reversal in its mode of predation. Seduction captures the not always equal ways in which subjects and objects reciprocally create each other through cosmic webs of predation.

A similar reversal occurs when the wife of a young man is pregnant. In Ávila such men are known as *aucashu yaya*, which means something like "fathers of beings that are not yet fully human" ("auca" refers to those people considered savages as well as to the unbaptized). Fetuses need continuous contributions of semen and the soul-stuff it contains in order to grow. As Hilario explained, "When the semen passes over" to the woman during sex, "the soul crosses too."[29] The resulting loss of soul-stuff over the course of a pregnancy weakens men. Rosalina once complained to her neighbor that her son had become extremely lazy and unable to hunt since his wife became pregnant. Her son had become soul blind to the other selves in the forest as a result of his soul loss. People in Ávila call this compromised condition *ahʰuas*. Expectant fathers experience morning sickness like their pregnant wives, and when the child is born they must observe a period of couvade through a variety of restrictions. They also become more aggressive throughout the pregnancy and are prone to fighting.

These expectant fathers lose their ability to be effective predators. They become soul blind. This is felt throughout the forest ecology of selves. Animals will suddenly refuse to enter the traps of expectant fathers, and when such men place fish poison in the water during communal fishing trips fish yields will be very low.

Game animals, recognizing this new status, no longer fear these hunters. Animals sense them as mean, and instead of becoming afraid of them they become angered and aggressive. What is more, even skittish herbivores begin to treat these once-formidable hunters as prey. Animals in the forest that are usually docile and wary, such as deer and the gray-necked wood rail (*pusara*), will suddenly become enraged and sometimes even attack these men. Ventura recounted to me that when his wife was pregnant deer in the forest suddenly charged him—on two separate occasions! And one of the deer even kicked him in the chest.

Ventura's sister, Angelicia, caught a baby coati in a spring trap and decided to keep it as a pet. Contemplating holding this creature in my arms, I asked her if the coati was liable to be aggressive toward me. Knowing that I was single, she laughed and then responded teasingly, "Only if you're an aucashu yaya . . ."

This weakened and soul blind condition of expectant fathers can be exploited. In the days when herds of white-lipped peccaries still passed through the Ávila region, hunters took the men into the forest and used them as charms, to attract these animals. As the peccaries—suddenly transformed into predators—would furiously charge the weakened and soul blind prey-victim, the victim's companions, who had been hiding in ambush, would jump out and kill the pigs.

Here again, through a process of seduction, predator and prey roles become reversed. The expectant father, unable to perceive other selves in the forest, has become an object. He is aicha—dead meat—to the peccaries and a tool, a charm, to his companions. Predator-prey relations are always nested, and this too is important for this charm to work. What at one level is a reversal of self-object relations (the expectant father is now hunted by his former prey) is nested within a higher-level relationship that reorients the direction of predation; the Runa—here a sort of distributed self in the figure of the group of hunters acting in unison—are reinstated as the true predator, and the pigs become meat, thanks to the temporarily desubjectified state of the expectant father.

Hunting charms in general attract animals that are considered "strong runners" (*sinchi puri*). These include tapirs, deer, and curassows. This too is in keeping with the idea that the goal of hunting and love charms is to make fully intentional selves come to men. The largely stationary and slow-moving sloths, by contrast, are not attracted by charms. Charms, then, are used with beings that are seen to have a lot of manifest "agency." Only very mobile beings—those with highly apparent intentionality—can be seduced. It is their agency, marked by their ability to act as if they were predators, that allows prey to be seduced. Game meat, aicha, must be alive before it can become dead.

In this regard, it is interesting to note that virtually all Ávila hunting and love charms come from animals.[30] There is, however, one notable exception: *buhyu panga*, a small hemiepiphytic vine belonging to the Araceae family.[31] It has the following unusual quality: when the torn pieces of its leaves are thrown into a stream, they dance around on top of the water's surface.[32] The name refers to the way the leaves' movements resemble those of pink river dolphins (*buhyu*) as they frolic in the confluences of rivers. Like the teeth of the river dolphin, this plant can become an ingredient for charms. Because the pieces of the leaves are drawn to each other and "stick together" (*llutarimun*) on the water's surface, this plant can attract game or women to the person who incorporates it into a charm. In general, hunting and love charms, in keeping with their purpose of effecting attraction, have as their ingredients only animal products because these come from organisms that are mobile. Buhyu panga, a leaf that moves on its own, is an exception that proves this rule.

Like predator/prey distinctions, gender functions as a shifting pronominal marker in this ecology of selves. When I was in the forest on hunting or plant collecting trips, my Runa companion would on many occasions detect game and then tell me to wait behind as he ran ahead with his gun cocked and ready to fire. Many times, as I waited quietly for him to return, the very game he was pursuing would approach me instead. I had this experience on several occasions. Troops of woolly monkeys high up in the canopy would circle back toward me. Capuchins would jump through the branches just above my head. Lone brocket deer would shoot past me, and small herds of collared peccaries would venture so close that I could almost touch them. When I asked why the animals would come to me instead of to the hunter the response was that, like a woman, I was unarmed and therefore the animals did not see me as a threatening predator and they were not frightened by my presence.

DEFAMILIARIZING THE HUMAN

Ethnographic fieldwork, involving intensive immersion in the lifeways—the language, the customs, the culture—of a foreign society, has traditionally been the preferred anthropological technique for critical self-reflection. Through an often painful and disorienting but ultimately liberating process, we immerse ourselves in a strange culture until its logics, meanings, and sentiments become familiar to us. By doing so, what we once took for granted—our natural and familiar way of doing things—comes, on our return home, to look strange. By stepping into another culture, fieldwork allows us, for a moment, to step outside of our own.

Anthropology allows us to move beyond our culture, but we never quite leave the human. What we are supposed to enter is always another culture. Ávila techniques of self-reflexive defamiliarization, Runa forms of anthropological wandering, by contrast, are not based on traveling to a different culture but on adopting a different kind of body. Natures are what become strange here, not cultures. Bodies are multiple and mutable, and the human body is only one of the many kinds of bodies that a self might inhabit. What kind of anthropology can emerge through this form of defamiliarizing the human?

Because eating entails such a palpable process of bodily transmutation, this form of reflexivity often involves ingestion. Some people in Ávila jokingly refer to edible leafcutter ants as people's crickets (*runa jiji*). Monkeys eat crickets, and when people eat ants—whole and sometimes even raw, crunchy exoskeleton and all—they too, in a certain sense, become monkeys. Another example: Many species of forest and cultivated trees belonging to the genus *Inga* (Fabaceae-Mimosoideae) are called *pacai* in Quichua. They produce edible fruits that can be pulled down off the tree and eaten. The flesh surrounding the seeds is fluffy, white, watery, and sweet. Another legume, *Parkia balslevii*, which belongs to the same subfamily, superficially resembles pacai in the shape of its fruits. The fruits of this tree are also edible, but its branches are very high and the fruits cannot be readily reached. Instead, they fall to the ground when they are overripe or rotten. The flesh begins to ferment and becomes brown and syrupy, like an off-flavored molasses. This tree is called *illahuanga pacai*, the vulture's pacai. From the perspective of vultures, rotting food is sweet; when the Runa eat vulture pacai, they adopt the point of view of a vulture; they come to enjoy rotting fruit as if it were fresh.

Seeing insects as appropriate food or seeing rotting things as sweet is something that other kinds of bodies do. When we eat ants-as-crickets or rotting vulture-pacai-as-sweet we are stepping out of our bodies into those of other beings, and in doing so, we see a *different* world from the subjective, I, point of view of another kind of embodiment. We are able, for a moment, to live in a different nature.

An inordinate interest in situating perspectives encourages an almost Zen-like mindfulness to one's precise state of being at any given moment. Here, as Luisa remembered them, are her exact thoughts at the precise moment her dogs were killed by a jaguar in the bush. The banality of her thoughts stands in marked contrast to the attack that was simultaneously taking place.[33]

> Here I was with my thoughts elsewhere,
> thinking, "should I go to Marina's or what?"
> With my mind somewhere else, thinking,
> "in order to go there
> I'll just quickly
> slip on a dress.
> But I no longer have a good dress to change into," I thought . . .

Luisa mindfully situates this daydream, and by extension herself, even though, as she says, she is not present but *elsewhere*. She locates herself in a "here" by mapping her thoughts to a different here: the site of the jaguar's attack on the dogs.

That attack occurred in the intimate female sphere of the abandoned gardens, a patchwork of transitional fallows and forests that Amériga, Delia, and Luisa would regularly frequent to collect fish poison, *chunda* palm fruits, and other products. By invading this domain, the jaguar had wandered outside of its proper territory deep in the forest. At one point Luisa angrily asked, "Are there no ridges at the banks of the Suno River?" "Ridges like that," she implored, "are the right places" for jaguars.[34] Because the jaguar that killed the dogs had undoubtedly been watching the women as they frequented their private gardens and fallows, Amériga, Delia, and Luisa were outraged. They felt that the presence of the jaguar in this intimate sphere was invasive. Delia noted that such places are supposed to be safe from predators. This is how América described the jaguar's violation of their intimate space:

> What kind of beast roams
> around our old dwellings
> just listening to us pissing?
> In those places where we've pissed, the jaguar's just walking around.

Imagining how one is seen in a very private moment through the eyes of another being is profoundly discomforting. It too is a form of defamiliarization, one that is highly disturbing, for it highlights the vulnerable nature of an isolated self, reduced to oneself—soul blind—cut off from others and exposed to a powerful predator.

SOUL BLINDNESS

What might it be like "see" ourselves in the very process of becoming blind to our own souls? One Ávila myth about the failed eradication of the *juri juri* demons, which Hilario related to his nephew Alejandro while sipping huayusa tea in the predawn hours, explores this terrifying possibility. This myth, I should note, parallels in a curious fashion the Spanish report of the 1578 uprising (see the introduction) in which all the Spaniards were killed, save, according to this account, a young girl who was spared because one of the natives wanted to marry her.

With the help of a tree lizard, the humans found the last hideout of the juri juri demons high up in a *chunchu* tree.[35] They ringed the tree with big piles of hot peppers, which they set on fire in order to choke out the demons. All the demons plummeted to their deaths except one. When this last juri juri finally fell to the ground she assumed the form of a beautiful white woman. A young man took pity on her. They married and began to raise a family. While bathing their children, the demon began to secretly eat them ("sucking their brains out, *tso tso*, from the crowns of their heads," América, to Hilario's annoyance, chimed in). One day the husband awoke from a magically induced sleep tormented by lice. He naively asked his wife to pick them out of his hair. She sat behind him, in a position that made her now invisible to him—a position that made it impossible for him to look back—and began combing her fingers through his hair. And then the man started to feel something strange.

His neck
became bu—rning hot[36]

He then observed, in a matter-of-fact way, detached from any emotion:

"I'm blee—ding
it would seem that
I'm wou- wounded"

And then, with a flat voice, devoid of any sentiment, the man concluded:

"you're eating me"

"It wasn't," Hilario explained, "like he was angry or anything." He was merely stating—"just like that"—the simple fact that he was being eaten alive.

And he just slept . . .
She made him sleep into his death.

The man is eaten alive but unable to experience this from a subjective perspective. He can never really "see" his wife, sitting behind him, eating him. He cannot return her gaze. Instead, he can only experience his own demise from an external disembodied stance. He can only logically deduce that he is wounded, and then that he is being eaten alive, by the physical effects this action produces. He has become completely "blind" to himself as a self. He feels no pain, nor does he suffer; he just registers the sensation that his neck is burning. Only later does he come to the realization that this is caused by his own blood flowing from his head. His demonic wife causes him to experience his death from outside his body. Before his life fades into indistinction—"Out of sleeping a waking, / Out of waking a sleep; / Life death overtaking; / Deep underneath deep?"—before he moves from affectless catatonia to sleep, and from sleep to death, he becomes an object to himself. He becomes inert, unfeeling. And his only awareness, however dimly perceived, is of this fact. This is a dystopian glimpse of a world where agency becomes divorced from a feeling, purposeful, thinking, embodied, and localized self. This is the final terminus of selfhood: radical soul blindness, an intimation of a world devoid of the enchantment of life, a world with no self, no souls, and no futures, just effects.

FOUR

Trans-Species Pidgins

> When *Thou* is spoken, the speaker has no thing for his object. For where there is a thing
> there is another thing. Every *It* is bounded by others; *It* exists only through being
> bounded by others. But when *Thou* is spoken, there is no thing, *Thou* has no bounds.
> When *Thou* is spoken, the speaker has no *thing*, he has indeed nothing. But he takes his
> stand in relation.
>
> —Martin Buber, *I and Thou*

The dogs should have known what was to befall them in the forest that day
they were killed. In a conversation she had with Delia and Luisa, back at the
house shortly after we buried the dogs' bodies, América wondered aloud why
her family's canine companions were unable to augur their own deaths and, by
extension, why she, their master, was caught unaware of the fate that would
befall them: "While I was by the fire, they didn't dream," she said. "They just
slept, those dogs, and they're usually real dreamers. Normally while sleeping
by the fire they'll bark, '*hua hua hua.*'" Dogs, I learned, dream, and by observing them as they dream people can know what their dreams mean. If, as
América suggested, their dogs would have barked "*hua hua*" in their sleep, this
would have been an indicator that they were dreaming of chasing animals, and
they would therefore have done the same in the forest the following day, for
this is how a dog barks when pursuing game. If, by contrast, they would have
barked "*cuai*" that night, this would be a sure signal that a jaguar would kill
them the following day, for this is how dogs cry out when attacked by felines.[1]

That night, however, the dogs didn't bark at all, and therefore, much to the
consternation of their masters, they failed to foretell their own deaths. As
Delia proclaimed, "Therefore, they shouldn't have died." The realization that
the system of dream interpretation that people use to understand their dogs

had failed provoked an epistemological crisis of sorts; the women began to question whether they could ever know anything. América, visibly frustrated, asked, "So how can we ever know?" Everyone laughed somewhat uneasily as Luisa reflected, "How is it knowable? Now, even when people are gonna die, we won't be able to know." América concluded simply, "It wasn't meant to be known."

The dreams and desires of dogs are, in principle, knowable, because all beings, and not just humans, engage with the world and with each other as selves, that is, as beings that have a point of view. To understand other kinds of selves, one simply needs to learn how to inhabit their variously embodied points of view. So the question of how dogs dream matters deeply. Not only because of the purported predictive power of dreams, but because imagining that the thoughts of dogs are not knowable would throw into question whether it is ever possible to know the intentions and goals of any kind of self.

Entertaining the viewpoints of other beings blurs the boundaries that separate kinds of selves. In their mutual attempts to live together and to make sense of one another, dogs and people, for example, increasingly come to partake in a sort of shared trans-species habitus that does not observe the distinctions we might otherwise make between nature and culture; specifically, the hierarchical relationship that unites the Runa and their dogs is based as much on the ways in which humans have been able to harness canine forms of social organization as it is on the legacies of a colonial history in the Upper Amazon that links people in Ávila to the white-mestizo world beyond their village.

Trans-species communication is dangerous business. It must be undertaken in ways that avoid, on the one hand, the complete transmutation of the human self—no one wants to permanently become a dog—and, on the other, the monadic isolation represented by what in the previous chapter I called soul blindness, which is the solipsistic flipside of this transmutation. To mitigate such dangers people in Ávila make strategic use of different trans-species communicative strategies. These strategies reveal something important about the need to venture beyond the human and the challenges of doing so in ways that don't dissolve the human. These strategies also reveal something important about the logic inherent to semiosis. Understanding these, in turn, is central to the anthropology beyond the human that I am developing. To tease out some of these properties, I've chosen, as a heuristic device to focus my inquiry, the following small but vexing ethnological conundrum: Why do people in Ávila interpret dog dreams literally (e.g., when a dog barks in its sleep this is an omen that it will bark in identical fashion the following day in the forest), whereas for the most part they interpret their own dreams metaphorically (e.g., if a man dreams of killing a chicken he will kill a game bird in the forest the following day)?

ALL TOO HUMAN

The ecology of selves within which the Runa, their dogs, and the many beings of the forest live reaches well beyond the human, but it is also one that is "all too human."[2] I use this term to refer to the ways in which our lives and those of others get caught up in the moral webs we humans spin. I wish to signal that an anthropology that seeks a more capacious understanding of the human by attending to our relations to those who stand beyond us must also understand such relations by virtue of the ways in which they can be affected by that which is distinctively human.

I argued in chapter 1 that symbolic reference is distinctively human. That is, the symbolic is something that is (on this planet) unique to humans. The moral is also distinctively human, because to think morally and to act ethically requires symbolic reference. It requires the ability to momentarily distance ourselves from the world and our actions in it to reflect on our possible modes of future conduct—conduct that we can deem potentially good for others that are not us. This distancing is achieved through symbolic reference.

My intention here is not to arrive at a universal understanding of what might be an appropriate moral system. Nor is it a claim that living well with others— what Haraway (2008: 288–89) calls "flourishing"—necessarily requires rational abstraction, or morality (even though thinking about the good does). But to imagine an anthropology beyond the human that does not simply project human qualities everywhere we must situate morality ontologically. That is, we must be precise about where and when morality comes to exist. To state it baldly, before humans walked this earth there was no morality and no ethics. Morality is not constitutive of the nonhuman beings with whom we share this planet. It is potentially appropriate to morally evaluate actions we humans initiate. This is not the case for nonhumans (see Deacon 1997: 219).

Value, by contrast, is intrinsic to the broader nonhuman living world because it is intrinsic to life. There are things that are good or bad for a living self and its potential for growth (see Deacon 2012: 25, 322), keeping in mind that by "growth" I mean the possibility to learn by experience (see chapter 2).

Because nonhuman living selves can grow it is appropriate to think about the moral implications our actions have on their potential to grow well—to flourish.[3]

As with the symbolic, to say that the moral is distinctive does not mean that it is cut off from that from which it emerges. Morality stands in a relation of emergent continuity to value, just as symbolic reference stands in a relation of emergent continuity to indexical reference. And value extends beyond the human. It is a constitutive feature of living selves. Our moral worlds can affect nonhuman beings precisely because there are things that are good or bad for them. And some of those things that are good or bad for them are also, we might learn if we could learn to listen to these beings with whom our lives are entangled, good or bad for us as well.

This is especially true when we begin to consider how this *us* that comprises us is an emergent self that can incorporate many kinds of beings in its coming configurations. We humans are the products of the multiple nonhuman beings that have come to make and continue to make us who we are. Our cells are, in a sense, themselves selves, and their organelles were once, in the distant past, free-living bacterial selves; our bodies are vast ecologies of selves (Margulis and Sagan 2002; McFall-Ngai et al. 2013). None of these selves in and of themselves are loci of moral action, even though larger selves with emergent properties (properties such as the capacity for moral thinking, in the case of humans) can subsume them.

The multispecies encounter is, as Haraway has intimated, a particularly important domain for cultivating an ethical practice. In it, we are most clearly confronted with what she calls "significant otherness" (Haraway 2003). In these encounters we are confronted by an otherness that is radically (significantly) other—without, I would add, that otherness being incommensurable or "incognizable" (see chapter 2). But in these encounters we can nonetheless find ways to enter intimate (significant) relations with these others who are radically not us. Many of these selves who are not ourselves are also not human. That is, they are not symbolic creatures (which means that they are also not loci of moral judgment). As such, they force us to find new ways to listen; they force us to think beyond our moral worlds in ways that can help us imagine and realize more just and better worlds.

A more capacious ethical practice, one that mindfully attends to finding ways of living in a world peopled by other selves, should come to be a feature of the possible worlds we imagine and seek to engender with other beings. Just how to go about doing this, just how to decide on what kind of flourishing to encourage—and to make room for the many deaths on which all flourishing depends—is itself a moral problem (see Haraway 2008: 157, 288). Morality is a constitutive feature of our human lives; it is one of human life's many difficulties. It is also something we can better understand through an anthropology beyond the human; semiosis and morality must be thought together because the moral cannot emerge without the symbolic.

The qualifier "all too" (as opposed to "distinctive") is not value-neutral. It carries its own moral judgment. It implies that there is something potentially troubling at play here. This chapter and those that follow attend to this by opening themselves to the complicated ways in which the Runa are immersed in the many all-too-human legacies of a colonial history that affect so much of life in this part of the Amazon. These chapters, in short, begin to open themselves to problems that involve power.

DOG-HUMAN ENTANGLEMENTS

In many ways dogs and people in Ávila live in independent worlds. People often ignore their dogs, and once they mature into adults their masters don't even necessarily feed them. Dogs, for their part, seem to largely ignore people. Resting in the cool shade under the house, stealing off after the bitch next door, or, as Hilario's dogs did a few days before they were killed, hunting down a deer on their own—dogs largely live their own lives.[4] And yet their lives are also intimately entangled with those of their human masters. This entanglement does not just involve the circumscribed context of the home or village. It is also the product of the interactions that dogs and people have with the biotic world of the forest as well as with the sociopolitical world beyond Ávila through which both species are linked by the legacy of a colonial history. Dog-human relationships need to be understood in terms of both these poles. The hierarchical structure on which these relationships are based is simultaneously (but not equally) a biological and a colonial fact. Relationships of predation, for example, characterize how the Runa and their dogs relate to the forest as well as to the world of whites.

Through a process that Brian Hare and others (2002) call "phylogenetic enculturation" dogs have penetrated human social worlds to such an extent that they exceed even chimpanzees in understanding certain aspects of human communication (such as different forms of pointing to indicate the location of food). Becoming human in the right ways is central to surviving as a dog in Ávila.[5] Accordingly, people strive to guide their dogs along this path in much the same way that they help youngsters to mature into adulthood. Just as they advise a child on how to live correctly, people counsel their dogs. To do this, they make them ingest a mixture of plants and other substances, such as agouti bile, known collectively as *tsita*. Some of the ingredients are hallucinogenic and also quite toxic.[6] By giving them advice in this fashion, people in Ávila are trying to reinforce a human ethos of comportment that dogs should share.[7]

Like Runa adults, dogs should not be lazy. For dogs, this means that instead of chasing chickens and other domestic animals, they should pursue forest game. In addition, dogs, like people, should not be violent. This means that dogs shouldn't bite people or bark at them loudly. Finally, dogs, like their masters, should not expend all their energy on sex. I've observed people administer tsita to dogs on several occasions. What happened at Ventura's house is typical in many respects. According to Ventura, before his dog Puntero discovered females he was a good hunter, but once he began to be sexually active he lost the ability to be aware of animals in the forest. Because soul-substance is passed to a developing fetus through semen during sex, he, like the expectant fathers I discussed in chapter 3, became soul blind. So early one morning Ventura and his family captured Puntero, fastened his snout shut with a strip of vine, and hog-tied him. Ventura then poured tsita down Puntero's snout. While doing this he said the following:

> chases little rodents
> it will not bite chickens
> chases swiftly
> it should say, "*hua hua*"
> it will not lie

The way Ventura spoke to his dog is extremely unusual. I'll return to it later. For now, I'll only give a general gloss. In the first phrase "little rodents" refers obliquely to the agoutis that dogs are supposed to chase. The second phrase is an admonition not to attack domestic animals but to hunt forest ones instead. The third phrase encourages the dog to chase animals but otherwise not to run ahead of the hunter. The fourth phrase reaffirms what a good dog should be doing: finding game and therefore barking "*hua hua*." The final phrase refers to the fact that some dogs "lie." That is, they bark "*hua hua*" even when there are no animals present.

As Ventura poured the liquid, Puntero attempted to bark. Because his snout was tied shut he was unable to do so. When he was finally released Puntero stumbled off and remained in a daze all day. Such a treatment carries real risks. Many dogs do not survive this ordeal, which highlights how dependent dogs are on exhibiting human qualities for their physical survival. There is no place in Runa society for dogs-as-animals.

Dogs, however, are not just animals-becoming-people. They can also acquire qualities of jaguars, the quintessential predators. Like jaguars, dogs are carnivorous. Their natural propensity (when they haven't succumbed to domestic laziness) is to hunt animals in the forest. Even when dogs are fed vegetal food, such as palm hearts, people in Ávila refer to it as meat in their presence.

People also see dogs as their potential predators. During the conquest the Spaniards used dogs to attack the forebears of the Ávila Runa.[8] Today this canine predatory nature is visible with regard to the special ritual meal that forms part of the feast known as the aya pichca, which I discussed in the previous chapter. This meal, which consists of cooked palm hearts, is eaten early in the morning after the ghost of the deceased is sent back to where he or she was born, to reunite with the afterbirth. The long tubular hearts, which are left intact for this meal, resemble human bones (by contrast, when palm hearts are prepared for everyday meals they are finely chopped).[9] Resembling bones, the palm hearts presented at this meal serve as a substitute for the corpse of the deceased in a sort of "mortuary endo-cannibalistic" feast, not unlike other feasts in other parts of Amazonia (and perhaps historically in the Ávila region as well; see Oberem 1980: 288) in which the bones of the dead are consumed by their living relatives (see Fausto 2007). Those present at the meal held after we sent off Jorge's ghost stressed that under no circumstances must dogs eat the palm hearts. Dogs, who see palm hearts as meat, are predators par excellence, for, like jaguars and cannibalistic humans they can come to treat people as prey.[10]

Dogs, then, can acquire jaguarlike attributes, but jaguars can also become canine. Despite their manifest role as predators, jaguars are also the subservient dogs of the spirit beings who are the masters of the animals in the forest. According to Ventura, "What we think of as a jaguar is actually a [spirit animal master's] dog."

It is important to note that in Ávila these spirit animal masters,[11] who keep jaguars as dogs, are often described as powerful white estate owners and priests. People liken the game animals these masters own and protect to the herds of cattle that whites keep on their ranches. In one sense, then, the Ávila

Runa are not so different from many other Amazonians who understand human and nonhuman sociality as one and the same thing. That is, for many Amazonians, the social principles found in human society are the same as those that structure animal and spirit societies of the forest. And this goes in both directions: nonhuman sociality informs understandings of human sociality just as much as human sociality informs that of nonhumans (see Descola 1994). Ávila, however, has always been part of larger political economies at the same time that it has been fully immersed in the forest's ecology of selves. This means that Runa "society" also includes a sense of the fraught relations the Runa have to others in a broader colonial, and now republican, arena. As a consequence, the sociality that extends to the nonhumans of the forest is also informed by those all-too-human histories in which the Runa, over the generations, have become entangled. This, then, in part, is why the animal masters that live deep in the forest are white (for a further discussion of what exactly being "white" here means, see chapters 5 and 6).

Were-jaguars—runa puma—are also dogs. As Ventura explained it to me, with reference to his recently deceased father, when a person "with jaguar" (*pumayu*) dies, his or her soul goes to the forest to "become a dog." Were-jaguars become the "dogs" of the spirit animal masters. That is, they become subservient to them in the same way that people from Ávila enter subservient relations when they go to work as field hands for estate owners and priests. A runa puma, then, is simultaneously Runa, a potent feline predator, and the obedient dog of a white animal master.

In addition to being emblematic of the Runa predicament of being simultaneously predator and prey, dominant and submissive, dogs are extensions of people's actions in the world beyond the village. Because they serve as scouts, often detecting prey well before their masters can, dogs extend Runa predatory endeavors in the forest. They are also, along with the humans, subject to the same threats of predation by jaguars.

In addition to the linkages they help people forge with the beings of the forest, dogs allow the Runa to reach out to that other world beyond the village—the realm of white-mestizo colonists who own ranches near Ávila territory. Ávila dogs are woefully underfed, and as a result they are often quite unhealthy. For this reason, they are rarely able to produce viable offspring, and people from Ávila must often turn to outsiders to obtain pups. A human-induced canine reproductive failure, then, makes people dependent on these outsiders for the procreation of their dogs. They also tend to adopt the dog names that colonists use. In this regard, the names Pucaña and Huiqui are exceptions. More common are dog names such as Marquesa, Quiteña, or even Tiwintza (a toponym of Jivaroan origin, marking the site of Ecuador's 1995 territorial conflict with Peru). This practice of using the dog names preferred by colonists is another indicator of how dogs always link the Runa to a broader social world, even when they are also products of a domestic sociability.

As a link between forest and outside worlds, dogs in many ways resemble the Runa, who, as "Christian Indians," have historically served as mediators between the urban world of whites and the sylvan one of the Auca, or non-Christian "unconquered" indigenous peoples, especially the Huaorani (Hudelson 1987; Taylor 1999: 195).[12] Until approximately the 1950s the Runa were actually enlisted by powerful estate owners—ironically, like the mastiffs of the Spanish conquest used to hunt down Runa forebears—to help them track down and attack Huaorani settlements.[13] And, as ranch hands, they continue to help colonists engage with the forest by, for example, hunting for them.

I should also note that the kinds of dogs that people in Ávila acquire from colonists do not for the most belong to any recognizable breed. Throughout much of Spanish-speaking Ecuador, such dogs are disparagingly described as "runa" (as in "un perro runa")—that is, as mutts. In Quichua, by contrast, runa means person. It is used as a sort of pronominal marker of the subject position—for all selves see themselves as persons—and it is only hypostasized as ethnonym in objectifying practices such as ethnography, racial discrimination, and identity politics (see chapter 6). This Quichua term for "person," however, has come to be used in Spanish to refer to mongrel dogs.[14] It would not be too far a stretch to suggest that runa for many Ecuadorians refers to those dogs that lack a kind of civilized status, those sin cultura, or without culture. Certain kinds of dogs and a certain group of indigenous people, the Quichua-speaking Runa, according to this colonial primitivist logic, have come to serve as markers along this imagined route from animality to humanity.

Trans-species relations often involve an important hierarchical component; humans and dogs are mutually constituted but in ways that are fundamentally unequal for the parties involved.[15] The domestication of dogs, beginning some fifteen thousand years ago (Savolainen et al. 2002), has been dependent, in part, on the fact that the progenitors of dogs were highly social animals that lived in well-established dominance hierarchies. Part of the process of domestication involved replacing the apex of this hierarchy in such a way that dogs would imprint on their human master as the new pack leader. Human-dog relations are dependent on the ways in which canine and human socialities merge, and they are always predicated, in some measure, on the ongoing establishment of relations of dominance and submission (Ellen 1999: 62). In colonial and postcolonial situations, such as that in which people in Ávila are immersed, this merger acquires renewed meaning. Dogs are submissive to their human masters in the same way that the Runa, historically, have been forced to be submissive to white estate owners, government officials, and priests (see Muratorio 1987). This position is not fixed, however. The lowland Runa, as opposed to some of their highland indigenous Quichua-speaking counterparts, have always maintained a relatively higher degree of autonomy vis-à-vis state authorities. They, and their canine companions, then, are also like powerful predatory jaguars that, for their part, are not just the servile dogs of the animal masters.

Adopting the viewpoint of another kind of being to a certain extent means that we "become" another kind "with" that being (see Haraway 2008: 4, 16–17). And yet these sorts of entanglements are dangerous. People in Ávila seek to avoid the state of monadic isolation that I've been calling soul blindness, by which they lose the ability to be aware of the other selves that inhabit the cosmos.[16] And yet they want to do so without fully dissolving that sort of selfhood distinctive to their position in this cosmos as human beings. Soul blindness and becoming an-other-with-an-other are opposite extremes along a continuum that spans the range of ways of inhabiting an ecology of selves. There is a constant tension, then, between the blurring of interspecies boundaries and maintaining difference, and the challenge is to find the semiotic means to productively sustain this tension without being pulled to either extreme.[17]

DREAMING

Because dreaming is a privileged mode of communication through which, via souls, contact among radically different kinds of beings becomes possible, it is an important site for this negotiation. According to people in Ávila, dreams are the product of the ambulations of the soul. During sleep, the soul separates from the body, its "owner,"[18] and interacts with the souls of other beings. Dreams are not commentaries on the world; they take place in it (see also Tedlock 1992).

The vast majority of dreams that are discussed in Ávila are about hunting or other forest encounters. Most are interpreted metaphorically and establish a correspondence between domestic and forest realms. For example, if a hunter dreams of killing a domestic pig he will kill a peccary in the forest the following day. The nocturnal encounter is one between two souls—that of the pig and that of the Runa hunter. Killing the pig's nocturnal domestic manifestation therefore renders soulless its forest manifestation to be encountered the following day. Now soul blind, this creature can be easily found in the forest and hunted because it is no longer cognizant of those other selves that might stand to it as predators.

Metaphoric dreams are ways of experiencing certain kinds of ecological connections among kinds of beings in such a manner that their differences are recognized and maintained without losing the possibility for communication. This is accomplished by virtue of the fact that metaphor is able to unite disparate but analogous, and therefore related, entities. It recognizes a gap as it points to a connection. Under normal waking circumstances, the Runa see peccaries in the forest as wild animals, even though they see them in their dreams as domestic pigs. But things get more complicated. The spirit animal masters who own and care for these animals (which appear as peccaries to the Runa in their waking lives) see them as their domestic pigs. So when people dream they come to see these animals from the spirit masters' point of view – as domestic pigs. Importantly, the spirit animal masters are considered dominant kinds of beings. From the perspective of these masters, the literal ground for the metaphoric relationship between peccary and domestic pig is the animal-as-domesticate. What is literal and what is metaphoric shifts. For the animal masters, what we would think of as "nature" (i.e., the "real" forest animals) is not the ground (cf. Strathern 1980: 189); peccaries are really domestic pigs. So one could say that from the perspective of an animal master, which is the dominant one and therefore the one that carries more weight, a hunter's dream of a pig is the literal ground for which his forest encounter with a peccary the following day will be a metaphor. In Ávila the literal refers to a customary interpretation of the world internal to a given domain. Metaphor, by contrast, is used to align the situated points of view of beings that inhabit different worlds. The distinction between figure and ground, then, can change according to context. What stays constant is that metaphor establishes a difference in perspective between kinds of beings inhabiting different domains. By linking the points of view of two beings at the same time that it recognizes the different worlds these beings inhabit, metaphor serves as a crucial brake that the Runa impose on the propensity toward blurring that is inherent to their way of interacting with other kinds of beings.

CANINE IMPERATIVES

Dreams, recall from the previous chapter, confirmed the identity of the predator that killed the dogs. Hilario's dead father's puma was the culprit. But Amériga's question remained unanswered. Why did the dogs fail to augur their own deaths? She felt that the dogs' dreams should have revealed the true nature of the forest encounter with the jaguar.

How could Amériga presume to know how her dogs dreamed? In order to address this, it is important to first understand in more detail how people in Ávila talk with their dogs. Talking to dogs is necessary but also dangerous; the Runa do not want to become dogs in the process. Certain modes of communication are important in this delicate cross-species negotiation, and it is to an analysis of these that I now turn.

It is due to their privileged position relative to animals in the trans-species interpretive hierarchy that constitutes the forest ecology of selves that the Runa feel they can readily understand the meanings of canine vocalizations.[19] Dogs, however, cannot, under normal circumstances, understand the full range of human speech. As I indicated earlier, if people want dogs to understand them they must give *the dogs* hallucinogenic drugs. That is, they must make their dogs into shamans so that they can traverse the boundaries that separate them from humans. I want to revisit in more detail the scene in which Ventura advised his dog on how to behave. While pouring the hallucinogenic mixture down Puntero's snout, he turned to him and said:

1.1 *ucucha-ta tiu tiu*
rodent-ACC chase[20]
chases little rodents[21]

1.2 *atalpa ama cani-nga*
chicken NEG IMP bite-3FUT
it will not bite chickens

1.3 *sinchi tiu tiu*
strong chase
chases swiftly

1.4 *"hua hua" ni-n*
"hua hua" say-3
it should say *"hua hua"* (the bark made when dogs are chasing animals)

1.5 *ama llulla-nga*
NEG IMP lie-3FUT
it will not lie (i.e., the dog should not bark as if it were chasing animals when
in reality it is not)

I am now in a position to explain why this is an extremely strange way of speaking.[22] When advising their dogs people in Ávila address them directly but in the third person. This appears to be similar to the Spanish *usted* system whereby third-person grammatical constructions are used in second-person pragmatic contexts to communicate status. Quichua, however, lacks such a deferential system. Notwithstanding, the Runa tweak Quichua to improvise one. That they are using grammatical constructions in new ways is most evident in line 1.2. In Quichua *ama* is typically used in second-person negative imperatives, as well as in negative subjunctives, but never in combination with the third-person future marker as it is being used here. I am dubbing this anomalous negative command a "canine imperative."[23]

Here is the challenge: in order for people to communicate with dogs, dogs must be treated as conscious human subjects (i.e., as *Yous*, even as *Thous*); yet dogs must simultaneously be treated as objects (*Its*) lest they talk back. This, it appears, is why Ventura uses this canine imperative to address Puntero obliquely.[24] And this also seems to be part of the reason that Puntero's snout was tied shut during this process. If dogs were to talk back, people would enter a canine subjectivity and therefore lose their privileged status as humans. By tying dogs down, in effect, denying them their animal bodies, they are permitting a human subjectivity to emerge. Canine imperatives, then, allow people to safely address this partially individuated emerging human self about the partially deindividuated and temporarily submerged canine one.[25]

The power-laden hierarchical relationship between dogs and humans that this attempt at communication reveals is analogous to that between humans and the spirit masters of animals. In the same way that people can understand their dogs, animal masters can readily understand the speech of humans; the Runa need only talk to them. Indeed, as I've observed on several occasions, in the forest people address these spirits directly. Under normal circumstances, however, humans cannot readily understand animal masters. Just as dogs require the hallucinogenic mixture tsita to understand the full range of human expression, people ingest hallucinogens, especially aya huasca, so that they can converse normally with these spirits. They use this opportunity to cement bonds of obligation with the spirit masters so that these, in turn, will allow them to hunt their animals. One important way of establishing such bonds is through the spirit master's daughters. Under the influence of hallucinogens, hunters attempt to cultivate amorous relations with them so that they will help them gain access to game meat via their fathers.

The relationship between these spirit lovers and Runa men is very similar to that between the Runa and their dogs. People give advice to their dogs in the third person and, in addition, tie their snouts shut, making it impossible for their dogs to respond. For related reasons, a spirit lover never allows her Runa partner to address her by name. Her proper name should be voiced only by other beings from the spirit master realm, and never in the presence of her human lovers. As one man told me, "One does not ask their names." Instead, men are only allowed to address their spirit lovers with the title *señora*. In Ávila this Spanish term is used to refer to and address white women regardless of marital status. By prohibiting Runa men from addressing them directly, the animal master's daughters can protect their privileged perspective as spirits and, in a sense, also as whites. This is analogous to the ways in which people communicate with their dogs so as to protect their own special position as humans.[26] At all levels, then, the goal is to be able to communicate across the boundaries that separate kinds without destabilizing them.

INTERSPECIES SPEECH

People use oblique forms of communication, such as canine imperatives, to put brakes on processes that threaten to blur the distinctions among kinds of beings. Yet the language that they use when talking to their dogs is simultaneously an instantiation of this same process of blurring. Accordingly, I have begun to think of it as a "trans-species pidgin." Like a pidgin it is characterized by reduced grammatical structure. It is not fully inflected, and it exhibits minimal clause embedding and simplified person marking. Furthermore, pidgins often emerge in colonial situations of contact. Given how in Ávila dog-human relations are entangled with Runa-white ones, this colonial valence seems particularly appropriate.

Indicative of its status as a trans-species pidgin, Runa dog talk—in a manner that is similar to the way Juanicu's were-jaguar compadre both spoke and panted (see chapter 3)—incorporates elements of communicative modalities from both human and animal realms. Using Quichua grammar, syntax, and lexicon, this "pidgin" exhibits elements of a human language. However, it also adopts elements of a preexisting trans-specific dog-human idiom. For example, *tiu tiu* (line 1.1) is used exclusively to spur dogs to chase game and is never used in human-human speech (except in quotation). In keeping with its paralinguistic identity, *tiu tiu* is not inflected here (see chapter 1). This interspecies pidgin also incorporates elements of dog talk. *Hua hua* (line 1.4) is an item from the canine lexicon. The Runa incorporate it into their utterances only through quotation. That is, they themselves would never bark. *Hua hua* is never inflected and is thus not fully integrated into human grammar. Both *tiu tiu* and *hua hua* involve reduplication, the iconic iteration of sound. This too is an important semiotic technique by which the Runa attempt to enter nonhuman, nonsymbolic referential modes.[27]

The Runa-dog trans-species pidgin is also like "motherese"—the purportedly distinctive form of language that adult caregivers use when speaking to babies—in that it exhibits grammatical simplification and is addressed to a subject that does not have full linguistic capabilities. This is an additional way in which it manifests a colonial valence. As we know, in many colonial and postcolonial contexts such as the Ávila one, natives come to be treated as standing to colonists as children stand to adults. Here is one example of how this plays out in Ávila. An engineer from the Ministry of Agriculture (Ministerio de Agricultura y Ganadería), visited Ávila, along with his wife and children, in order to confer on it the legal status of "personhood" (*personería jurídica*) as a state-recognized indigenous community (*comuna*). A number of people told me that he had come to give them "advice," for which they used the verb *camachina*—a term that is also used to describe how adults "counsel" children and dogs. In his conversations with me, the engineer, in turn, referred to the inhabitants of Ávila, regardless of age, as *"los jóvenes"* (youths, children). He and his wife—who, fittingly, is a schoolteacher—considered it their civic duty to mold the Ávila Runa into proper (i.e., mature, adult) Ecuadorian citizens. In fact, they insisted on beginning the annual communal meeting with the national anthem, and they spent much of the long meeting reading and explaining portions of the Ecuadorian constitution and carefully guiding the villagers through the government-mandated guidelines for democratically electing the comuna leaders. With titles such as president, vice president, treasurer, and secretary, these leaders would, ideally, simultaneously reproduce the bureaucratic apparatus of the state in the microcosm of the community and serve as the link between the village and the state. As I explore in the final chapter of this book, the contours of the self in Ávila are as much the product of the relations people have with nonhumans as they are the product of these sorts of intimate (and often paternalistic) encounters through which a larger nation-state comes to be manifested in their lives.

THE CONSTRAINTS OF FORM

The human-canine trans-species pidgin, like motherese, is oriented toward beings whose linguistic capabilities are in question. Although people in Ávila go to great lengths to make their dogs understand human speech, how they communicate with their dogs must also conform to the exigencies of those species that cannot normally understand human speech, with its heavily symbolic mode of reference. My cousin Vanessa who accompanied me on the unpleasant bus trip over the Andes into the Oriente (see chapter 1), finally got to visit Ávila with me. Not long after arriving at Hilario's house, however, she had the misfortune of being bitten on the calf by a young dog. The next afternoon, this dog, herself a fresh arrival (having been recently brought by one of Hilario's sons from across the Suno River where this son works as a field hand for colonists), bit her again. Hilario's family was quite disturbed by this behavior—the dog's "humanity" was at stake and, by extension, that of her masters—and Hilario and his other son Lucio therefore gave the dog the hallucinogenic tsita mixture and proceeded to "give her advice" in much the same way that Ventura had counseled Puntero. On this occasion, however, they took the drugged dog, with her mouth securely tied, and placed her snout against the same spot where she had bitten Vanessa the day before. While they were doing this Hilario said:

5.1 *amu amu mana canina*
[She, Vanessa, is a] master, a master and is not to be bitten

5.2 *amu amu amu imapata caparin*
[She is a] master, a master, a master, and there is no reason to bark

5.3 *amuta ama caninga*
It will not bite the master

Here, as visible in line 5.3, Hilario employs the same negative "canine imperative" construction that Ventura used. On this occasion, however, this phrase and the series of utterances in which it is embedded are entangled with an earnest nonlinguistic and nonsymbolic effort at communication with the dog. Whereas the negative canine imperative—"it will not bite"—responds to the challenge of speaking to the dog in such a way that, under the influence of hallucinogens, she can understand but not respond, the reenactment of the act of biting Vanessa serves as another form of negative canine imperative, here, however, not in a symbolic register but in an indexical one. As such, it responds to a different but equally important challenge: how to say "don't" without language.

Regarding this challenge of how to say "don't" without language, Bateson noted an interesting feature of communication visible among many mammals, including dogs. Their "play" employs a kind of paradox. When, for example, dogs play together they act as if they are fighting. They bite each other but in ways that are not painful. "The playful nip," observed Bateson (2000e: 180), "denotes the bite, but it does not denote what would be denoted by the bite." There is a curious logic at work here. It is as if, he continues, these animals were saying, "'These actions in which we now engage do not denote what those actions *for which they stand* would denote'" (180). Thinking of this semiotically, and here I follow Deacon (1997: 403–5), whereas negation is relatively simple to communicate in a symbolic register, it is quite difficult to do so in the index- ical communicative modalities typical of nonhuman communication. How do you tell a dog not to bite when the only secure modes of communication avail- able are via likeness and contiguity? How do you negate a resemblance or a relation of contiguity without stepping outside of strictly iconic and indexical forms of reference? Saying "don't" symbolically is simple. Because the symbolic realm has a level of detachment from indexical and iconic chains of semiotic associations it easily lends itself to meta-statements of this sort. That is, via symbolic modalities it is relatively easy to negate a statement at a "higher" interpretive level. But how do you say "don't" indexically? The only way to do so is to re-create the "indexical" sign but this time without its indexical effect. The only way to indexically convey the pragmatic negative canine imperative, "Don't bite" (or, in its Runa trans-species pidgin deferential form, "It will not bite"), is to reproduce the act of biting but in a way that is detached from its usual indexical associations. The playful dog nips. This "bite" is an index of a real bite, but it is so in a paradoxical way. Although it is an index of a real bite and all its real effects, it also forces a break in an otherwise transitive indexical chain. Because of the absence of a bite, a new relational space emerges, which we can call "play." The nip is an index of a bite but not an index of what that bite is itself an index. By re-creating the attack on my cousin, Hilario and Lucio attempted to enter into this canine play logic, constrained as it is, by the formal properties characteristic of indexical reference. They forced the dog to bite Vanessa again but this time with her snout tied shut. Theirs was an attempt to rupture the indexical link between the bite and its implications, and in this way to tell their dog "don't" through the idiom of a trans-species pidgin that, for the moment, has gone well beyond language.

It is never entirely clear whether and to what extent animals can understand human speech. If dogs could readily understand humans there would be no need to give them hallucinogens. The point I wish to make is that trans-species pidgins really are middle grounds (sensu White 1991; see also Conklin and Graham 1995). It is not enough to imagine how animals speak, or to attribute human speech to them. We are also confronted by, and forced to respond to, the constraints imposed by the particular characteristics of the semiotic modalities animals use to communicate among themselves. Regardless of its success, this attempt reveals a sensitivity on the part of people in Ávila to the formal constraints (see Deacon 2003) of a nonsymbolic semiotic modality.

THE CONUNDRUM

I want to return for a moment to the discussion, from this book's introduction, taken up again in the previous chapter, of the admonition to never look away from a jaguar encountered in the forest. Returning the jaguar's gaze encourages this creature to treat you as an equal predator—a *You*, a *Thou*. If you look away, it may well treat you as prey, soon-to-be dead meat, an *It*. Here too, in this nonlinguistic exchange, status is conveyed across species lines through the use of either direct or oblique modes of nonlinguistic communication. This too is a parameter of the zone in which canine imperatives operate. Jaguars and humans, then, enjoy a sort of parity according to people in Ávila. They can potentially entertain each other's gaze in a trans-species but nevertheless, to some extent at least, intersubjective space. For this reason some people maintain that if they eat lots of hot peppers they can repulse the jaguars they might encounter in the forest because eye contact will burn the jaguar's eyes. By contrast, eye contact with beings of higher levels is prohibitively dangerous. One should, for example, avoid such contact with the demons (*supaiguna*) that wander the forest. Looking at them causes death; by entertaining their gaze one enters their realm—that of the nonliving.[28]

In Ávila this sort of hierarchy of perspective is reflected in modes of communication. Literal communication takes place when one being can entertain the subjective viewpoint of the other. "Higher" beings can readily do this vis-à-vis lower ones, as is evident by the fact that people can understand dog "talk" or that spirits can hear the supplications of people. "Lower" ones, however, can only see the world from the perspective of higher beings via privileged vehicles of communication, such as hallucinogens, which can permit contact among souls of beings inhabiting different realms. Without special vehicles of communication, such as hallucinogens, lower beings understand higher ones only through metaphor, that is, through an idiom that establishes connections at the same time that it differentiates.

We can now address the conundrum with which I began this chapter: if metaphor is so important in Runa dreams and in other situations in which the differences between kinds of beings are recognized, why do the Runa interpret the dreams of their dogs literally?

In a metaphoric human dream people recognize a gap between their mode of perception and that of the animal masters. Through dreaming, they are able to see how the forest really is—as the domestic gardens and fallows of the dominant animal masters. This, however, is always juxtaposed to how they see the forest in their waking life—as wild. People in Ávila interpret dog dreams literally because they are able to see directly the manifestations of how their dogs' souls experience events thanks to the privileged status that they enjoy vis-à-vis dogs. By contrast, regarding the oneiric ambulations of their own souls, which involve interactions with dominant beings and the animals under their control, humans do not usually enjoy this privileged perspective. And this is why their dreams exhibit a metaphoric gap.

TRANS-SPECIES PIDGINS

In dog dream interpretation the gaps that separate kinds of beings, gaps that are often assiduously respected, collapse, at least for a moment, as dogs and people come together as part of a single affective field that transcends their boundaries as species—as they come together, in effect, as an emergent and highly ephemeral self distributed over two bodies.[29] Amériga's epistemic crisis reveals the tenuous nature but also the stakes of such a project. Dog dreams do not belong only to dogs. They are also part of the goals, fears, and aspirations of the Runa—the dogs' masters and occasional 'cosmonautical' companions— as they reach out, through the souls of their dogs, to engage with the beings that inhabit the world of the forest and beyond.

The sorts of entanglements I have discussed in this chapter are more than cultural, and yet they are not exactly noncultural either. They are everywhere biological, but they are not just about bodies. Dogs really become human (biologically and in historically specific ways) and the Runa really become puma; the need to survive encounters with feline semiotic selves requires it. These processes of "becoming with" others change what it means to be alive; and they change what it means to be human just as much as they change what it means to be a dog or even a predator.

We must be attendant to the danger-fraught, provisional, and highly tenuous attempts at communication—in short, the politics—involved in the interactions among different kinds of selves that inhabit very different, and often unequal, positions. Such attempts are inextricably tied up with questions of power. It is because *Thou* can be spoken when addressing dogs that dogs must, at times, be tied up: "Every *It* is bounded by others." Negotiating this tension between *It* and *Thou* that is inherent to living with others is a constant problem as people in Ávila struggle to take a stand "in relation" to the many kinds of other beings that inhabit their cosmos.

Runa-dog trans-species pidgins do more than iconically incorporate dog barks, and they do more than invent new human grammars adequate to this risky task of speaking in a way that can be heard across species lines without invoking a response. They also conform to something more abstract about the referential possibilities available to any kind of self, regardless of its status as human, organic, or even terrestrial,[30] and this involves the constraints of certain kinds of semiotic forms. When Hilario attempted to say "don't" without language he could only do so in one way. He and his dog fell into a form—one that is instantiated in but also sustains and exceeds not only the human but also the animal. It is toward an analysis of these sorts of forms, how they permeate life, how, given the appropriate constraints, they so effortlessly propagate across radically different kinds of domains, and how they come to acquire a peculiar social efficacy that I turn in the next chapter.

FIVE

Form's Effortless Efficacy

It is the people who are outside of the monastery who feel its atmosphere. Those who are practicing actually do not feel anything.

—Shunryu Suzuki, *Zen Mind, Beginner's Mind*

One night, while staying at Ventura's house, I dreamed I stood outside of a pen on a large cattle ranch like the one that belongs to a burly colonist, located just beyond Ávila territory on the way to Loreto. Inside, a collared peccary was running around. Suddenly, it stopped right in front of me. We both just stood there, looking at each other. Our intimacy overwhelmed me with a strange and novel feeling, an unexpected sense of resonance with this distant creature. I had an epiphany. I grasped something. I discovered, I think, a kind of love for that pig. But I also wanted to kill it. After some fumbling with a broken gun I had borrowed from one of the villagers I finally managed to shoot it point-blank. I cradled its limp body in my arms and went back to Ventura's house, proud that I would now have plenty of meat to share with his family.

What I dreamed that night is entwined with something that had happened the day before as Ventura and I were returning from a walk in the forest. Ventura sensed something and motioned for me to wait quietly while he ran up ahead to investigate, cocked gun at the ready. As I waited a collared peccary approached me. We both froze, our eyes fixed on each other, before it ran off.

This experience and its oneiric reverberation captured something about a moment of personal intimacy with a forest being and some of the contradictions that hunting such beings implies. People in Ávila, like many others who live in close contact with nonhuman beings, recognize many animals as potential persons with whom, on occasion, they have "personal" interactions (see Smuts 2001). My forest encounter with the peccary that afternoon, however fleeting, was an intimation of the possibility for this kind of trans-species intimacy. It served as a reminder that animals, like us, are selves; they represent the world in certain ways and act on the basis of those representations (see chapter 2). Yet hunting requires both recognizing this and treating these singular selves as generic objects; its goal, after all, is to transform them into pieces of meat for consumption and exchange (see chapter 3).

Ventura's take on my dream, however, didn't emphasize the tension I felt between recognizing animals as selves and the subsequent desubjectivization that killing them requires. As an experienced hunter Ventura was already adept at negotiating this. Instead, he was interested in what this dream had to say about my relationship to the animal's master—the spirit who owns the pig. Such masters of the beings of the forest are often thought of as European priests or powerful white estate owners, like that colonist, with his defiant swagger, pickup truck, and pigpen, who lives along the way to Loreto.

These spirit masters are a part of everyday life in Ávila. Ventura himself entered their realm when as a child he got lost in the forest. Accompanied by his dog, he was out hunting with his father. As the day wore on Ventura lagged farther and farther behind until boy and dog lost their way. He eventually met a girl he thought was his sister and followed her down a road that seemed to be taking them home but instead led them through a waterfall to the abode of the masters. After a few days, Ávila shamans, who were able to enter the spirit realm with the help of the hallucinogen aya huasca, managed to negotiate Ventura's release. By this time, however, he and his dog had become feral or wild (*quita* in Quichua). They lost the ability to recognize Ávila villagers as people. The dog failed to bark when called to, and Ventura didn't recognize, and was even frightened of, his own mother, Rosa.

Decades later, during the time of my stay in Ávila, Ventura's mother, now quite elderly and easily confused, also wound up in the realm of the spirit masters. One day, while caring for some of her grandchildren, Rosa simply wandered off into the forest. A full five weeks after her disappearance a young woman, fishing with her little brother in the forest, stumbled on her by a stream after first noting that the fish had been scared off by some presence. Rosa survived—emaciated, her scalp and toes worm-infested—long enough to report how a boy she took to be one of her teenage grandsons led her to the underground city of the masters that she called "Quito." This subterranean city, she said, was beautiful and opulent, "just like the living Quito," Ecuador's Andean capital.

I never expected to experience this master realm personally. But, according to Ventura, this is exactly what had happened. That I had dreamed of the peccary inside a pen, he explained, indicated that it was the spirit master of the animals that had allowed me to share in that intimate moment of mutual trans-species recognition the day before. The pig belonged to the spirit master of the forest, and the pen in which I saw it was on that master's ranch.

In juxtaposing a certain kind of human sociality with a wild one, my dream was a lot like one Juanicu's son Adelmo had. Early one morning Adelmo bolted out of bed and announced loudly, "I've dreamed!" before grabbing his shotgun and rushing out of the house. He returned a few hours later carrying a peccary over his shoulders. When I asked him what had prompted him to run out like that he replied that he had dreamed of buying a pair of shoes. The shoe stores in Loreto, filled with shelves of shoes and piles of rubber boots, provide an apt image for the profusion of tracks left by a herd of peccaries at a mud wallow. Furthermore, those smelly omnivorous pigs are social beings but not exactly in ways that the Runa would deem appropriate. In this regard, they are like one of those Lycra-clad colonist shopkeepers (revealing parts of their bodies in ways that no one in Ávila would). They are also like the "naked" Huaorani, the longtime "wild" enemies of the "civilized" (and clothed) Runa.[1]

My dream also shared something in common with one Fabian, a young father of two, had while we were out at his hunting camp. His was of a well-stocked general store filled with things like sacks of rice and cans of sardines and tended by a young priest. He later explained that this dream augured killing woolly monkeys. These monkeys travel in troops deep in the mountains, far from Runa settlements. Once spotted, they are relatively easy to hunt— usually several can be taken—and they are coveted for their thick layers of fat. Like the deep forests that these monkeys frequent the well-stocked general stores are at some distance from Runa settlements. And, like the monkey troops, the stores offer a cornucopia of food. Both the store and the monkey troops are controlled by powerful whites. Given the proper means, the Runa can have access to some of the wealth of both.

Dreams reflect a widespread Amazonian way of seeing human and nonhuman sociality as continuous with each other in a manner that also posits a rigorous parallel between human domestic realms and nonhuman sylvan ones (see Descola 1994). The game birds that the Runa encounter in the forest are really the chickens of the spirit forest masters, just as jaguars are the master's hunting and guard dogs.

What we humans see as wild, then, is, from the dominant perspective of the masters, domestic (see chapter 4). In contrast to our Euro-American multiculturalism, which assumes a uniform nature and multiple and variable culturally situated representations of it, this Amazonian understanding of the forest and its beings is something more akin to what Viveiros de Castro (1998) calls a multinatural one (see chapter 2). There exist many different natures, the products of the bodily dispositions of the different kinds of beings that inhabit the universe. But there is only one culture—an *I* perspective that all selves, human and nonhuman alike, inhabit. Culture in this sense is an *I* perspective. That is, from their *I* perspectives all beings see the different natures they inhabit as cultural: a jaguar—as an *I*—sees peccary blood as the manioc beer that is the customary staple of the Runa diet, and spirits, according to this same logic, see the forest as an orchard.

Why this echoing between cultural and natural, domestic and wild? And why should I be privy to it? This is not something multinaturalism can address; an anthropology beyond the human can. One might think that the ways in which this special kind of doubling logic infected my dreams is the by-product of sustained ethnographic fieldwork, a sort of enculturation to which eager anthropologists might feel susceptible. Except, as I've already hinted, culture may not be the best marker of difference in these parts of the world. In fact, as I hope the following discussion will illuminate, and following the argumentation of chapter 2, difference may not be the right starting point for understanding the broader problem of relating to which my dream gestures.

Moreover, I wasn't the only outsider to have experienced these resonances. I've since discovered that several missionaries and explorers passing through the region have also, apparently spontaneously, become attuned to these same sorts of parallels between human and forest realms. For instance, the nineteenth-century British explorer Alfred Simson, who stayed briefly in a Runa village, described Britain to a man named Marcelino in a way that unwittingly re-created the realm of the spirit masters of the forest. He matched up, through a series of isomorphic relations, the urban, opulent, domestic, and white realm of Britain, on the one hand, with the sylvan, impoverished, wild, and Indian one of the Amazon, on the other. Instead of villages scattered through the forest, there are large cities, he explained, and in place of scarcity, "knives, axes, beads ... and all such things were to be had there in the greatest profusion." In his country, he continued, instead of wild beasts there are only useful and edible ones (Simson 1880: 392–93).[2]

The conversation between Simson and Marcelino also hinted at shamanistic attempts to commensurate these realms. When the Runa die they go to live forever in the realm of the spirit masters, and so it is fitting that Simson refers to Britain as a "paradise." Access to this realm required an arduous journey that, according to Simson, might last some "ten moons"—a journey that, we later learn, Marcelino understood as being of the shamanistic sort. As they spoke Simson offered one of his pipes of "strong tobacco," and Marcelino proceeded to swallow "all the smoke he could draw in huge volumes" (1880: 393).

Tobacco, along with the hallucinogen aya huasca, is one of the vehicles that help people enter the point of view of the masters. In fact, people in Ávila refer to shamans as those "with tobacco" (*tabacuyu*). And thanks to the privileged access to other points of view that dreaming provides, I too, like Marcelino and the aya huasca-drinking Ávila shamans that rescued Ventura and his dog, was able to see the forest as it really is. I came to see it as a domestic space—a ranch—because this is how it appears from the dominant *I* perspective of the spirit master of the forest who owned the pig.

Why should such a parallel between sylvan and domestic—ecology and economy—appear in so many places, including my dreams? And why would a place like Quito come to be located deep in the forest? The claim I wish to make in this chapter is that addressing these seemingly disparate questions requires understanding something that might not, on the surface, seem relevant: it requires understanding the peculiar characteristics of regularities, habits, or patterns. In more abstract terms, I am arguing that getting at these questions requires an understanding of how certain configurations of constraint on possibility emerge and of the particular manner in which such configurations propagate in the world in ways that result in a sort of pattern. That is, addressing these questions requires understanding something about what I call "form."

The point I will be fleshing out is this: what encourages Amazonian forest ecologies and human economies to be aligned in my dreams and in those of the Runa is the pattern or form that such systems share. And this form, I wish to stress, is the result of something other than the imposition of human cognitive schema or cultural categories onto these systems.

It is hard to broach the topic of form beyond the human, as I do here, without being accused of making a Platonic argument for the separate existence of a transcendent realm of, say, ideal triangles or squares. By contrast, it is less controversial to consider the role form plays within the realm of the human.

The human mind, we can all agree, traffics in generalities, abstractions, and categories. Another way to say this is that form is central to human thought. Let me rephrase this statement in terms of the definition of *form* I have proposed: constraints on possibility emerge with our distinctively human ways of thinking, which result in a pattern that I here call form. For example, the associational logic of symbolic reference (treated in chapter 1 and revisited later in this chapter), which is so central to human thought and language, results in the creation of general concepts, such as, say, the word *bird.*

Such a general concept is more constrained than the various actual utterances of the word *bird* through which it is instantiated. Utterances, then, are more variable, less constrained, and "messier" than the concept they express. That is, there will be great variation in how any particular utterance of a word such as *bird* actually sounds. And yet the general concept, to which all of these particular utterances refer, allows these many variable utterances to be interpreted as meaningful instantiations of the concept "bird." This general concept (sometimes termed a "type") is more regular, more redundant, simpler, more abstract, and, ultimately, more patterned than the utterances (referred to as "tokens" in their relation to such types) that instantiate it. Thinking of such concepts in terms of form gets at this characteristic generality that a type exhibits.

Because language, with its symbolic properties, is distinctively human, it is all too easy to relegate such formal phenomena to human minds. And this encourages us to take a nominalist position. It encourages us to think of form solely as something humans impose on a world otherwise devoid of pattern, category, or generality. (And if we are anthropologists it encourages us to search for the origins of such categories in the distinctively human historically contingent, changing social and cultural contexts in which we are immersed; see chapter 1.) But taking such a position would be tantamount to allowing human language to colonize our thinking (see introduction and chapters 1 and 2). Given that, as I argued in previous chapters, human language is nested within a broader representational field made up of semiotic processes that emerge in and circulate in the nonhuman living world, projecting language onto this nonhuman world blinds us to these other representational modalities and their characteristics.

The human, then, is only one source of form. It is important to note, for the argument at hand, that an important characteristic that these semiotic modalities that exist beyond the human exhibit is that they too possess formal properties. That is, as with symbolic representation, these semiotic modalities (those made up of icons and indices) also exhibit constraints on possibility that result in a certain pattern.

I alluded to this at the end of the previous chapter in my discussion of the limited ways in which one can attempt to "say" "don't" in a nonsymbolic, non-linguistic, register and how the logic of this formal constraint on possibility is also manifested in a pattern of nonhuman animal communication—a form—that is visible in animal "play." That this pattern recurs time and again in many different species, and even in attempts at communication that cross species lines, is an example of the emergence and circulation of form in the world beyond the human.

As I mentioned in chapter 1, that semiosis exists beyond human minds and the contexts they create is one indicator that "generals," that is, habits, or regularities, or, in Peircean terms, "thirds," are "real." (By "real" here, I mean that such generals can come to manifest themselves in ways that are independent of humans, and they can come to have eventual effects in the world.) However—and this is key—whereas semiosis is in and of the living world beyond the human, form emerges from and is part and parcel of the nonliving one as well.

That is, form is a sort of general real despite the fact that it is neither alive nor a kind of thought. This can be hard to appreciate given the ways in which life and thought harness form and are everywhere made over by its logics and properties. So in this chapter I am taking anthropology a step further beyond the human to explore the way in which a particular manifestation of a general exists in the world beyond life.

Throughout this book, especially in chapter 1, I have been discussing a number of generals. Emergent phenomena are generals. Habits or regularities are generals. All of these, in some way or another, are the result of constraints on possibility (see Deacon 2012). I am using the term *form* to refer to the particular manifestations of the generals I treat here. I do so to emphasize some of the geometrical patternings involved in the ways generals become expressed in the Amazon. Many of these could be classed as self-organizing emergent phenomena, or in Deacon's (2006, 2012) terms, "morphodynamic"—that is, characterized by dynamics that generate form (see chapter 1).

Examples of such nonliving emergent forms in the Amazon include, as I will discuss, the patterned distribution of rivers or the recurrent circular shapes of the whirlpools that sometimes form in them. Each of these nonliving forms is the product of constraints on possibility. Regarding rivers, water doesn't just flow anywhere in the Amazon. Rather, the distribution of rivers is constrained by a variety of factors, which results in a pattern. Regarding whirlpools, under the right conditions swift currents moving around obstructions create self-reinforcing circular patterns that are a subset of all the possible (messier, less constrained, more turbulent) ways in which water might otherwise flow.

In recognizing the emergence of form in the physical world, then, this chapter requires an excursion beyond the living. The goal, however, is to see what it is that the living "do" with form and the particular ways in which what they do with it is infected by form's strange logics and properties. As I will show, humans in the Amazon harness such forms, and so do other kinds of living beings.

Form, then, is crucial to lives, human and otherwise. Nevertheless, the workings of this vague entity remain largely undertheorized in anthropological analysis. This, in large part, is due to the fact that form lacks the tangibility of a standard ethnographic object. Nevertheless, form, like the basic intentionality of the pig and the palpable materiality of its meat, is something real. Indeed, its particular mode of efficacy will require us to think again what we mean by the "real." If, as anthropologists, we can find ways to attend ethnographically to those processes of form amplification and harnessing as they play out in the Amazon, we might be able to become better attuned to the strange ways in which form moves through us. This, in turn, can help us harness form's logics and properties as a conceptual tool that might even help us rethink our very idea of what it means to think.

RUBBER

To get a better handle on form, I'd like to turn to another forest/city juxtaposition, not unlike Rosa's Quito-in-the-forest or Marcelino's Britain. Manuela Carneiro da Cunha (1998) has described how a Jaminaua shamanic novitiate of the Juruá River system of Amazonian Brazil traveled vast distances downriver to apprentice in the port cities on the Amazon itself, in order to be recognized as a powerful shaman upon returning to his village. To understand why these port cities have come to be the conduits for indigenous shamanic empowerment, we need to understand something of a momentous period in Amazonian history: the rubber boom, which began in the late nineteenth century and reached into the second decade of the twentieth, and the particular kinds of isomorphic correspondences that made this boom possible in the first place.

In many respects the rubber boom that swept through the Amazon was the product of a variety of techno-scientific, "natural-cultural," and imperial conjunctures. That is, the discovery of vulcanization coupled with the invention and mass production of automobiles and other machines catapulted rubber onto an international market. For the Upper Amazon this boom was a sort of second conquest, given that outsiders were dependent, for the most part, on exploiting local populations to extract this increasingly valuable commodity that was dispersed throughout the forest. The boom, however, ended abruptly after rubber seedlings, which had been removed from the Amazon basin by British naturalists, began to take hold in Southeast Asian plantations (see Brockway 1979; Hemming 1987; Dean 1987). This story, told in terms of such interactions among humans, and even among human and nonhuman beings, is well known. Here, I want to discuss something not often noticed: namely, the ways in which the peculiar properties of form mediated all these interactions and made this extractive economic system possible.

Let me explain what I mean. Rubber falls into a form. That is, there is a specific configuration of constraints on the possible distribution of rubber trees. The distribution of rubber trees throughout the Amazon forests—whether the preferred *Hevea brasiliensis* or a few other latex-producing taxa—conforms to a specific pattern: individual rubber trees are widely dispersed throughout the forest across vast stretches of the landscape. Plant species that are widely dispersed stand a better chance of surviving attacks from species-specific pathogens,[3] such as, in the case of *H. brasiliensis*, the fungal parasite *Microcyclus ulei*, which causes the disease known as South American leaf blight. Because this parasite is endemic throughout rubber's natural range, rubber could not be easily cultivated in high-density plantations there (Dean 1987: 53–86). An interaction with this parasite results in a particular pattern of rubber distribution. Individual rubber trees are, for the most part, widely and evenly distributed and not clumped in single-species stands. The result is that rubber "explores," or comes to occupy, landscape in a way that manifests a specific pattern. Any attempt to exploit rubber in situ must recognize this.[4]

The distribution of water throughout the Amazonian landscape also conforms to a specific pattern or form. This has a variety of causes. Due to a number of global climactic, geographic, and biological factors, there is a lot of water in the Amazon basin. Furthermore, water only flows in one direction: downhill. Thus small creeks flow into larger streams, which in turn flow into small rivers that flow into larger ones, and this pattern repeats itself until the enormous Amazon disgorges into the Atlantic Ocean (see figure 1, on page 4).

For largely unrelated reasons there exist, then, two patterns or forms: the distribution of rubber throughout the landscape and the distribution of waterways. These regularities happen to explore landscape in the same way. Therefore, wherever there is a rubber tree it is likely that nearby there will be a stream that leads to a river.

Because these patterns happen to explore landscape in the same way, following one can lead to the other. The Amazon rubber economy exploited and relied on the similarities these patterns share. By navigating up the river network to find rubber and then floating the rubber downstream, it linked these patterns such that these physical and biological domains became united in an economic system that exploited them thanks to the formal similarities they share.

Humans are not the only ones who link floristic and riverine distribution patterns. The fish known in Ávila as *quiruyu*,[5] for example, eats fruits of the aptly named tree *quiruyu huapa*[6] when these fall into rivers. This fish, in effect, uses rivers to get at this resource. In doing so it also potentially propagates the patterned similarities—the form—that floristic and riverine distributions share. If in eating these fruits the fish were to disperse its seeds along the course of the river, then the pattern of this plant's distribution would come to match that of the rivers even more closely.

The Amazon riverine network exhibits an additional regularity crucial to the way rubber was harnessed via form: self-similarity across scale. That is, the branching of creeks is like the branching of streams, which is like the branching of rivers. As such, it resembles the compound ferns that people in Ávila call *chichinda*, which also exhibit self-similarity across scale. *Chinda* refers to a haphazard pile, especially to a tangled mass of driftwood such as the kind that might snag around the base of a riverbank tree after a flood. By reduplicating a part of this word—*chi-chinda*—this plant name captures how in a compound fern the pattern of divisions of the frond at one level is the same as that of the next higher-order level of divisions. *Chichinda*, which alludes to a tangled mass nested within another tangled mass, captures this fern's self-similarity across scale; a pattern at one level is nested within the same pattern at a higher more inclusive one.

The river network's self-similarity is also unidirectional. Smaller rivers flow into larger ones, and water becomes increasingly concentrated across an ever smaller expanse of landscape as one moves down the hydrographic network. Da Cunha (1998: 10–11) has highlighted a curious phenomenon in the Juruá River basin during the rubber boom period. A vast network of creditor-debt relations emerged, which assumed a nested self-similar repeating pattern across scale that was isomorphic with the river network. A rubber merchant located at one confluence of rivers extended credit upriver and was in turn in debt to the more powerful merchant located downriver at the next confluence. This nested pattern linked indigenous communities in the deepest forests to rubber barons at the mouth of the Amazon and even in Europe.

Humans, however, are not the only ones who harness the unidirectionally nested riverine pattern. Amazon river dolphins, like traders, also congregate at the confluences of rivers (Emmons 1990; McGuire and Winemiller 1998). They feed on the fish that accumulate there due to this nested characteristic of the river network.

Being inside form is effortless. Its causal logic is in this sense quite different from the push-and-pull logic we usually associate with the physical effort needed to do something. Rubber floated downstream will eventually get to the port. And yet a great amount of work was required to get rubber into this form. It took great skill and effort to find the trees, extract and then prepare the latex into bundles, and then carry these to the nearest stream.[7] More to the point, it took great coercive force to get others to do these things. During the rubber boom, Ávila, like many other Upper Amazonian villages, was raided by rubber bosses looking for slave labor (Oberem 1980: 117; Reeve 1988).

It is not surprising that villages such as Ávila should attract the attention of rubber bosses, for their inhabitants were already adept at harnessing forest forms to get at resources. Just as rubber tapping involves harnessing the riverine form to get at trees, hunting involves harnessing form. Because of the high species diversity and local rarity of species and the lack of any one fruiting season, the fruits that animals eat are highly dispersed in both space and time (Schaik, Terborgh, and Wright 1993). This means that at any given time there will exist a different geometrical constellation of fruiting resources that attracts animals. Fruit-eating animals amplify this constellation's pattern. For they are not only attracted to fruiting trees but often also to the increased safety provided by foraging in a multispecies association. Each member "contributes" its species-specific abilities to detect predators—resulting in a greater overall group awareness of potential danger (Terborgh 1990; Heymann and Buchanan-Smith 2000: esp. 181). That predators, in turn, are attracted to this concentration of animals further amplifies the pattern of distribution of life across the forest landscape. This results in a particular pattern of potential game meat: a clustered, shifting, highly ephemeral and localized concentration of animals interspersed by vast areas of relative emptiness. Ávila hunters, then, don't hunt animals directly. Rather, they seek to discover and harness the ephemeral form created by the particular spatial distribution or configuration

FIGURE 7. Rubber-boom–era hunters-of-hunters. Courtesy of the Whiffen Collection, Museum of Archeology and Anthropology, Cambridge University.

of those tree species that are fruiting at any given point in time because this is what attracts animals.[8]

Hunters, those already adept at harnessing forest forms, make ideal rubber tappers. But to get them to do this often meant hunting these hunters like animals. Rubber bosses often enlisted members of hostile indigenous groups to do it. In an image reproduced by Michael Taussig (1987: 48) of such hunters-of-hunters in the Putumayo region of the Colombian Amazon it is no coincidence that the man in the foreground is wearing jaguar canines and white clothing (see figure 7).

By adopting the bodily habitus of a predatory jaguar and a dominant white (a classic multinatural perspectival shamanistic strategy; see chapter 2), he can come to see the Indians he hunts as both prey and underlings. Those hunters-of-hunters that Taussig writes about were referred to as "muchachos"—boys—a reminder of the fact that they too were subservient to someone else: the white bosses. The rubber economy amplified an existing hierarchical trophic pattern of predation (with carnivores, such as jaguars, "above" the herbivores, such as deer, that they prey on), and in the process this economy combined this pattern with a paternalistic colonial one.

Ávila, as I mentioned, was by no means safe from slave raiding. In fact, one of the first stories Amériga told me on my initial trip to Ávila in 1992 was of how, as a child, her own grandmother was spared from slavery by being unceremoniously pushed out through the back bamboo wall of her house just as the raiders arrived at the front door. Ávila, in the foothills of the Andes, is far away from the navigable rivers and high-quality sources of rubber. *Hevea brasiliensis*, which produces the best rubber, doesn't grow near Ávila. Nevertheless, through great coercive effort many Ávila inhabitants were pushed into the rubber economy's form. They were forcibly relocated far downriver on the Napo in what is now Peru, and even beyond, where navigable rivers and rubber trees were abundant. Almost none returned.[9]

The rubber boom economy was able to exist and grow because it united a series of partially overlapping forms, such as predatory chains, plant and animal spatial configurations, and hydrographic networks, by linking the similarities these share. The result was that all these more basic regularities came to be part of an overarching form—an exploitative political-economic structure whose grasp was very difficult to escape.

In fact, this form created the conditions of possibility for the political relations that emerged. Shamans, those experts at stepping into dominant points of view within a multinatural perspectival system of cosmic predation, harnessed it to gain power. By apprenticing downriver the Jaminaua shaman was able to adopt a perspective that encompassed and exceeded the viewpoints of the social actors upstream (da Cunha 1998: 12). Being downriver means inhabiting a more inclusive level of the river pattern's nested self-similarity—a form that had now become socially important thanks to a colonial economy that linked it to the forest and to its indigenous inhabitants.[10] What is more, Amazonian shamanism cannot be understood outside of the colonial hierarchy that in part created it and to which it responds (see Gow 1996; Taussig 1987). However, shamanism is not just a product of colonialism. Shamanism and colonial extraction are equally caught up, constrained by, and forced to harness a shared form that partially exceeds them.

EMERGENT FORMS

Forms, such as the pattern that brings rubber trees, rivers, and economies into relation with one another, are emergent. By "emergent," I don't just mean new or indeterminate or complex. Rather, with reference to my discussion in chapter 1, I mean the appearance of unprecedented relational properties, which are not reducible to any of the more basic component parts that give rise to them.

Form, as an emergent property, makes itself manifest in the physical landscape of the Amazon. Take, for example, whirlpools, such as those that sometimes arise in Amazonian rivers, which I discussed earlier in this book and in the introduction to this chapter. Such whirlpools possess novel properties with respect to the rivers in which they appear; namely, they come to exhibit a coordinated circular pattern of moving water. This circular pattern in which the water in a whirlpool flows is more constrained and thus simpler than the otherwise freer, more turbulent, and hence less patterned flow of water in the rest of the river.

The whirlpool's circular form emerges from the river's water, and this is a phenomenon that cannot be reduced to the contingent histories that give that water its specific characteristics. Let me explain. Any given unit of water flowing through the Amazonian watershed certainly has a particular history associated with it. That is, it is, in a sense, affected by its past. It flowed through a particular landscape and it acquired different attributes as a result. Such histories—where the water came from, what happened to it there— certainly give different Amazonian rivers their specific characteristics. If, for example, the water flowing into a particular river passed through nutrient-poor white-sand soils, that river's water would become tannin-rich (see chapter 2), and hence dark, translucent, and acidic. However, and crucially for the argument at hand, such histories do not explain or predict the form the whirlpool will take in such rivers. Under the right conditions a circular shape will emerge regardless of the particular histories of where the water in the rivers came from.

Importantly, however, the conditions that result in the emergence of a whirlpool include the continuous flow of water. So the novel form a whirlpool takes is never fully separable from the water from which it emerges: block the river's flow, and the form will disappear.

And yet the whirlpool is something other than the continuous flow, which it requires. That something other is also something less. And this "something less" is why it makes sense to think of emergent entities such as a whirlpool in terms of form. As I mentioned, water flowing through a whirlpool does so in a way that is less free when compared to all the various less coordinated ways in which water otherwise moves through a river. This redundancy—this something less—is what results in the circular pattern of flow we associate with whirlpools. It is what accounts for its form.

In being both different from and continuous with that from which they come, and on which they depend, whirlpools are like other emergent phenomena, such as, for example, symbolic reference. Symbolic reference, recall from chapter 1, emerges out of those other more basic semiotic modalities within which it is nested. Like a whirlpool and its relationship to the water flowing in the river, symbolic reference exhibits new emergent properties with respect to the icons and indices on which it depends and from which it comes.

This characteristic of disjuncture-despite-continuity that appears with whirlpools also applies to the emergent pattern visible in the rubber economy. The disparate causes responsible for rubber and river distributions become irrelevant once an economic system unites them by virtue of the regularities that rubber and rivers share. And yet such an economy is everywhere, obviously, dependent on rubber. And it is also dependent on the rivers used to access that rubber.

Emergent phenomena, then, are nested. They enjoy a level of detachment from the lower order processes out of which they arise. And yet their existence is dependent on lower-order conditions. This goes in one direction: whirlpools disappear when riverbed conditions change, but riverbeds do not depend on whirlpools for their persistence. Similarly, the Amazon rubber economy was wholly dependent for its existence on the ways in which parasites such as the South American leaf blight constrained rubber's distribution. Once rubber plantations in Southeast Asia—far removed from these parasites—began to produce latex, this crucial constraint responsible for the patterned distribution of rubber trees disappeared. An entirely different economic arrangement became possible, and, like a fleeting whirlpool, the emergent form, the political-economic system that united rubber, rivers, natives, and bosses, vanished.

The biosocial efficacy of form lies in part in the way it both exceeds and is continuous with its component parts. It is continuous in the sense that emergent patterns are always connected to lower-level energetics and materialities. And the materialities—say, fish, meat, fruits, or rubber—are what living selves, be they dolphins, hunters, fruit-eating fish, or rubber bosses, are trying to access when they harness form. Form also exceeds these in the sense that as these patterns become linked their similarities propagate across very different kinds of domains: the regularities through which rubber is harnessed cross from the physical to the biological to the human.

In this process by which forms come to be combined at higher levels, however, the higher-order emergent pattern also acquires properties specific to antecedent ones. The rubber boom economy was nested like the rivers and predatory like portions of the tropical food chain. It captured something of these other-than-human forms. But it also integrated them into an emergent form that is, in addition, all too human (see chapter 4). Let me explain. The nonhuman forms I've been discussing here—those, for example, that involve nesting and predation—are hierarchical without being moral. It makes no sense to downplay the importance of hierarchical forms in the nonhuman world. This is not the way to ground our moral thinking, because such forms are not in any way moral. Hierarchy takes on a moral aspect in all-too-human worlds only because morality is an emergent property of the symbolic semiosis distinctive to humans (see chapter 4). Although themselves beyond the moral and hence amoral (i.e., nonmoral), such hierarchical patterns nonetheless get caught up in systems with all-too-human emergent properties—systems such as the highly exploitative economy based on rubber extraction, whose moral valence is not reducible to the more basic formal alignments of hierarchical patterns on which it depended.

THE MASTERS OF THE FOREST

But why, returning to Ávila and my dream, is it that the realm of the spirit masters unites hunting in the forest with the larger political economy and colonial history in which the Runa are also immersed? What, in short, does it mean to say that these spirit masters of the forest are also "white"?

Whiteness is just one element in a series of partially overlapping hierarchical correspondences that are superimposed in the spirit realm of the masters of the forest. For instance, each mountain around Ávila is owned and controlled by a different spirit master. The most powerful of these lives in an underground "Quito" located inside Sumaco Volcano, the region's highest peak. This volcano also lends its name to the early-sixteenth-century jurisdiction, the *provincia de Sumaco*, in recognition of the paramount chief to whom all regional subchiefs owed allegiance before this area succumbed to colonial rule and came to be known by the Spanish name Ávila.[11] Lesser forest masters live in cities and villages that are likened to the smaller towns and cities that make up the parish and provincial seats of Ecuador's Amazonian provinces. These correspond to the region's smaller mountains. The masters living in these stand in the same relation to the master living in the underground Quito as the pre-Hispanic and early colonial subchiefs stood to the paramount chief associated with Sumaco Volcano.

This mapping of pre-Hispanic and contemporary administrative hierarchies onto a topographic one partially overlaps with the network of estates or haciendas that dominated the local extractive resource economy until recent times and articulated that economy to Quito. The realm of the spirit masters is also a bustling productive estate, like the great rubber-boom–era haciendas along the Napo River.[12] And the masters travel to and from their pastures and fallows, shuttling game animals in their pickup trucks and airplanes. Hilario, who many years ago climbed to the top of Sumaco Volcano with a crew of army engineers intent on erecting a relay antenna there, reported that the gullies he saw emanating radially from its sugar cone summit are the highways of the masters. In the same way that roads originate in Quito and from there extend throughout Ecuador, all the major rivers of the greater Ávila region originate from this mountaintop.

It is my contention that the realm of the spirit masters superimposes ethnic, pre-Hispanic, colonial, and postcolonial hierarchies on the landscape because all of these various sociopolitical arrangements are subject to similar constraints regarding how certain biotic resources can be mobilized across space. That is, if Amazonian household economies and broader national and even global ones attempt to capture bits of the living wealth that the forest houses—whether in the form of game, rubber, or other floristic products—they can only do so by accessing the conjunction of physical and biotic patternings in which this wealth is caught up.[13] As I've mentioned, hunters, for the most part, don't hunt animals directly; they harness the forms that attract animals. In a similar fashion, estate owners, through debt peonage, and during certain periods even outright enslavement, collected forest products via the Runa. This extractive pattern creates a clustered distribution. Like the pattern of fruiting trees that attracts animals, haciendas came to be nodes where forest resources and the city ones with which they were commensurated became concentrated. It is the hacienda that harbored the "greatest profusion" of "knives, axes, and beads" (Simson 1880: 392–93), and it is the hacienda that accumulated the forest products that the Runa, in turn, exchanged for these. Cities, like Quito, also exhibit this clumped pattern of wealth accumulation, insofar as these are both the sources of trade goods and the end points for forest products.

The lowland Runa had an intimate and yet fraught relationship to Quito and its wealth. They were sometimes charged with the task of carrying whites on their backs to this city (Muratorio 1987). And in the days when Ávila was considerably more isolated from markets its inhabitants would go directly to Quito, making the eight-day trek, along with their forest products, in the hope of exchanging their goods for some of the wealth that the city harbored.

In the higher-order emergent realm of the spirit masters of the forest, hunting, estates, and cities align with each other by virtue of the similarities they share regarding their relations to the patterns of resource distribution that exist around them. Hierarchy is crucial to form propagation across these different domains. Spirit realms unite these various overlapping forms at a "higher" emergent level in the same way that the rubber economy is at a "higher" level than the patterns of rubber trees and rivers it unites. How form is amplified in human domains clearly is the contingent product of all-too-human histories. And yet hierarchy itself is also a kind of form, which has unique properties that exceed the contingencies of earthly bodies and histories, even if it is only instantiated in these.[14]

SEMIOTIC HIERARCHY

This interplay between the logical formal properties of hierarchy and the contingent ways in which it comes to acquire a moral valence is visible in those transspecies pidgins, discussed in the previous chapter, through which the Runa attempt to understand and communicate with other beings. The hierarchy involved in trans-species communication clearly has a colonial inflection; that's why I call them pidgins. As discussed in chapter 4, dogs, for example, often occupy the same structural position vis-à-vis the Runa as the Runa do vis-à-vis whites. Recall that although some Runa turn into powerful jaguars when they die, as jaguars they also become the dogs of the white spirit masters. These sorts of colonial hierarchies, however, are the morally loaded emergent amplifications of more fundamental nonhuman ones that are devoid of any moral valence.

Many of these more fundamental hierarchies involve the nested and unidirectional properties inherent to semiosis. To recapitulate from chapter 1, and to further develop something I alluded to above, symbolic reference, that distinctively human semiotic modality, which is based on conventional signs, has

emergent semiotic properties with respect to the more basic iconic and indexical referential strategies (i.e., those that involve signs of likeness and contiguity, respectively) that we humans share with all other life-forms. These three
representational modalities are hierarchically nested and connected. Indices,
which form the basis for communication in the biological world, are the product of higher-order relations among icons, and as such they have novel, emergent referential properties with respect to icons. Similarly, symbols are the
product of higher-order relations among indices, also with novel emergent
properties with respect to indices. This only goes in one direction. Symbolic
reference requires indices, but indexical reference does not need symbols.

These emergent hierarchical properties that make human language (based
as it is on symbolic reference) a distinctive semiotic modality also structure the
ways in which people in Ávila differentiate between animal and human realms.
Let me illustrate this with an exchange that took place between Luisa, Delia,
Amériga, and a squirrel cuckoo. This exchange took place not long after the
family's dog Huiqui had returned from the forest badly wounded by a jaguar.
This example shows the role that hierarchy plays, particularly as it structures
the perceived distinction between levels of meanings in different semiotic registers. Animal vocalizations, taken at face value as "utterances," are at one level
of significance, whereas the more general "human" messages that these vocalizations might also contain can emerge at another, higher level.

When the exchange in question took place the women had just returned
from collecting fish poison in their transitional orchards and fallows. They
were at home, sipping beer, peeling manioc, and still uncertain about the fate
of the other two dogs. We had not yet gone out to search for them and did not
yet know that they had been killed by a jaguar, although at this point this is
what the women thought had happened, and that scenario provided the interpretive frame for the conversation they were having.

As the women talked they were abruptly interrupted by a squirrel cuckoo
calling *"shicuá'"* as it flew over the house. Immediately afterward, Luisa and
Amériga simultaneously interjected the following:

Luisa:	Amériga:
shicuá'	*"Shicúhua,"* it says

The squirrel cuckoo, known in Ávila as *shicúhua*, has a variable call. If you
hear it calling *"ti' ti' ti,"* as people in Ávila imitate one of its vocalizations, it is
said to be "speaking well" and what you are at the moment desiring will come

to pass. However, if you should hear it making the call we heard that day as the bird flew overhead, a vocalization that people in Ávila imitate as "*shicuá*," that which you think will happen will not come to be and the bird is therefore said to be "lying." Other animals, I should note, call in similar ways. The pygmy anteater, known by the related name *shicúhua indillama*, makes an ominous hiss that portends that a relative will die.

Importantly, however, neither this hiss nor the squirrel cuckoo's call *shicuá'* in and of itself is a prophetic sign. Rather, although these vocalizations can certainly be treated on their own as signs, they only acquire their particular significance as a sort of omen when they are interpreted to be manifestations of the Quichua word *Shicúhua*. The word *Shicúhua*, pronounced with attention to the tendency in Quichua toward penultimate stress, not the cuckoo's squawk *shicuá'* or the pygmy anteater's hiss, is what causes these otherwise meaningful vocalizations to be treated, in addition, as portents.

This difference between the squirrel cuckoo's squawk *shicuá'* and *Shicúhua*, which is what this bird is said to be "saying" in making this vocalization, is important. As the squirrel cuckoo flew overhead Luisa imitated its call as she heard it: "*shicuá'*." Amériga, by contrast, quoted it: "'*Shicúhua*,' it says." In the process, Amériga also pronounced the call in a way that was less faithful to the sound the bird actually made and more in keeping with the stress patterns in Quichua.[15]

Whereas Luisa imitated what she heard, and thus constrained herself to the utterance as instance, Amériga tried to get at what the bird was "saying" more generally. She was in effect interpreting the message within "human language," which, I should note, is the literal meaning of *runa shimi*, the Quichua name for what the Runa speak. As such, she treated it as standing to the "token" animal utterance as a "type." Let me illustrate by virtue of an English example. In English any particular utterance of, say, "bird" is taken as an instantiation—or token—of the word *Bird*, which stands to it as a general concept—or type. My point is that something similar is going on here. Amériga treated the squirrel cuckoo's squawk as an instantiation of a sort of species-specific token of the "human" word *Shicúhua*, which stands to this squawk as a type. And just as we can interpret any utterance of "bird" by virtue of its relation to the word *Bird* in English, so too Amériga interpreted this animal vocalization as being an instance of a more general "human" word, *Shicúhua*. As such this vocalization is now understood to carry a particular message. Species-specific vocalizations (whether the squirrel cuckoo's squawk or the pygmy anteater's hiss) can act as individual tokens of more general terms in the "human" language Quichua, which serve as their types.

I want to emphasize that it is not that the squawk in and of itself is necessarily meaningless; it can still be interpreted by humans (and others) as an indexical sign. But it acquires its additional meaning as a particular kind of omen in a specific divinatory system when it is seen as being an instance of something more general.

To treat this squawk as meaningful at this level—to treat it as an omen— Amériga brought the squirrel cuckoo's call into language. The squawk *shicuá'* became legible as an instantiation of *Shicúhua*. Understood as a manifestation of "human language" this call (which might otherwise be indexically meaningful) now carried with it an additional prophetic message in a symbolic register.

And the women acted on this. The operative assumption that until now had been guiding the conversation—that the dogs had been killed—was now, it seemed, wrong. Amériga, accordingly, reinterpreted the dogs' plight within the new framework of assumptions suggested by the call. Heeding the cuckoo's message, she now imagined an alternative scenario that would explain why the dogs hadn't come home yet: "Having eaten a coati," she conjectured, "they're out there wandering around with their bellies full."[16] Delia wondered how then to account for the puncture wound on the head of the dog that straggled home. "So what happened?" she asked.[17] After a short pause Amériga suggested that on being attacked perhaps the coati bit the dog. Thanks to the kind of call the squirrel cuckoo made, and the system through which the women interpreted it, Amériga, Luisa, and Delia began to hope that the dogs had not encountered a feline but had instead simply scrapped with a coati and were still alive.

One might rightly say that this particular system of omens that I have been describing is specific to humans, or that it is specific even to a particular culture. And yet distinguishing between animal tokens and their human types as the women were doing is something more than a human (or cultural) imposition onto "nature." This is because the distinctions they make draw on the formal hierarchical properties that distinguish symbols from indices. These formal semiotic properties, which are neither innate nor conventional nor necessarily human, confer on human symbolic reference some of its unique representational characteristics when compared to the semiosis that is more generally distributed throughout the biological world. While indices point to instances, symbols have a more general application since their indexical power is distributed throughout the symbolic system in which they are immersed. Yet symbols represent in a manner that draws on indices in special ways (see Peirce CP 2.249). This is visible in the distinction people in Ávila make between *shicuá'* and *Shicúhua*. *Shicuá'*, a token animal vocalization, which can otherwise be simply interpreted indexically (signifying the presence of a bird, of danger, etc.), can be understood to carry an additional message when it is interpreted as an instantiation of the more general human word *Shicúhua* that stands to it as a type. That type gains traction in the world by virtue of its token manifestations.

In short, the difference between how Luisa and América treated the squirrel cuckoo's call reveals a hierarchical (i.e., unidirectional, nested) distinction between the not-necessarily-human semiosis of life and a human form of semiosis that takes up this nonhuman semiosis in special ways. This distinction between these two kinds of semiosis is neither biological nor cultural nor human; it is formal.

THE PLAY OF FORM

In locating manifestations of type/token distinctions in Runa attempts to make sense of the forest's semiosis, I've been discussing hierarchy-as-form. But I want to pause for a moment to reflect on the possibilities inherent to another kind of form propagation, also manifest in these trans-species pidgins, which is less hierarchical and more lateral or "rhizomatic." Later that afternoon, long after América's interpretation of the squirrel cuckoo's call changed the conversation's direction—long after we discovered that, this shift in direction notwithstanding, the dogs had indeed been killed by a jaguar—América and Luisa recalled how as they collected fish poison out in the brush they each heard the spot winged antbird call. The spot winged antbird calls *"chíriqui,"* as people in Ávila imitate it, when jaguars startle them. This call is therefore a well-known indicator of the presence of jaguars, and it is also the onomatopoeic source for *chiriquíhua*, which is the name for this bird in Ávila.

Back at the house, América and Luisa simultaneously reflected on how from their respective positions in the brush they heard this antbird call at the moment of the attack:

América:

shina manchararinga
that's how it gets scared

Luisa:

paririhua paririhua
from heliconia to heliconia

runata ricusa
even seeing a person

shuma' shuma'
from one to another

manchana
it gets scared

chíriqui' chíriqui'

"*Chiriquíhua Chiriquíhua,*" nin
saying, "*Chiriquíhua Chiriquíhua*"

chi uyararca
that's what could be heard

"*-quíhua*"

imachari
what might it mean?

In their parallel recollections of this event Amériga pronounced the bird's name and sought its meaning. The bird was "saying, '*Chiriquíhua*'" (and not simply calling *chíriqui'*). And because its utterance now conformed to the systemic norms of a general and pan-cosmic runa shimi, what it said now surely had some sort of ominous meaning, even though what exactly this implied Amériga wasn't at the time quite sure.

Luisa, by contrast, simply imitated what "could be heard" and allowed this to resonate with other sonic images:

paririhua paririhua
shuma' shuma'
chíriqui' chíriqui'

Hers was an image of the antbird startled by a jaguar, flitting nervously throughout the underbrush from one heliconia leaf to another. Translating liberally, one gets an image of this bird going from

leaf to leaf
jumping jumping
chíriqui' chíriqui'

Freed from the interpretive drive to stabilize the call's meaning, Luisa was able to trace the bird's ecological embedding through a kind of play that is open to the possibilities inherent to the iconic propagation of sonic form. Ignoring for a moment the ways in which "*chíriqui'*" might refer "up" to *Chiriquíhua*—a word that "means" something in a broader, relatively more fixed symbolic system—allowing it simply to resonate with other images and tracing out these relations, has, then, its own "significant" possibilities.

I want to emphasize the point that eschewing a certain kind of stabilized meaning does not make Luisa's exploration nonsemiotic. "*Chíriqui'*" is meaningful without necessarily meaning something. It has a different truck with significance—one that, relatively speaking, is more iconic in logic. Amériga, by contrast, was attempting to extract information from the antbird's call. Surely semiosis serves to convey what Bateson termed "the difference which makes a difference" (see chapter 2), but, as Luisa's reaction to the antbird indicates, focusing only on how representational systems convey difference misses something fundamental about the ways in which semiosis also depends on the effortless propagation of form. Iconicity is central to this.

In this regard, I want to return to my discussion in chapter 1 of those cryptically camouflaged Amazonian insects known in English as walking sticks and referred to by entomologists as phasmids. I want to think about these insects here in terms of form. Their iconicity, as I mentioned, is not based on someone out there noticing that they look like twigs. Rather, the walking stick's likeness is the product of the fact that the ancestors of its potential predators did not notice the differences between their ancestors and actual twigs. Over evolutionary time those walking stick lineages that were least noticed survived. In this way a certain form—the "fit" between twig and insect—came to propagate effortlessly into the future.

Form, then, is not imposed from above; it falls out. This, of course, is an outcome of a kind of interpretive effort that is more intuitively familiar to us; it results from the ways in which predators "work" to notice the differences between certain insects and their environments. These are the insects that, not being twiglike enough, are eaten. The relation that iconicity has to confusion or indifference (see chapter 2), as the proliferation of "twigginess" reveals, gets at some of the strange logic of form and its effortless propagation.

As Luisa's verbal play illustrates, iconicity has a certain freedom from our limiting intentions. It can leap out of the symbolic—but not out of semiosis or significance. Given the right conditions it can effortlessly explore the world in ways that can create unexpected associations.

This kind of exploratory freedom is I think what Claude Lévi-Strauss (1966: 219) was getting at when he wrote of savage thought (not to be confused with the thought of "savages") as "mind in its untamed state as distinct from mind cultivated or domesticated for the purpose of yielding a return." It is also something, I believe, that Sigmund Freud grasped in his recognition of how the unconscious partakes of the kind of self-organizing logic to which Lévi-Strauss is alluding. Such a logic is well exemplified in Freud's (1999) writing on dreams. It is also visible in his treatment of slips of the tongue, malapropisms, and forgotten names. These emerge in the course of everyday conversation when for some reason the intended word is repressed (Freud 1965) and they sometimes, as Freud noted with wonder, circulate contagiously from one person to another (85). English translations of his work call these "mistaken" utterances parapraxes, from *parapraxia*, the defective performance of certain purposive acts. That is, when thought's "purpose of yielding a return" is removed what is left is that which is ancillary to or beyond what is practical: the fragile but effortless iconic propagation of self-organizing thought, which resonates with and thereby explores its environment. In the case of parapraxis this can take the form of the spontaneous production of alliterative chains that link a forgotten word to a repressed thought (Freud 1965: 85). Freud's insight, gesturing quite literally to an "ecology of mind," was to develop ways to become aware of these iconic associative chains of thought (and even to find ways to encourage them to proliferate) and then, by observing them, to learn something about the inner forests these thoughts explore as they resonate through the psyche.

Freud, of course, wanted to tame this kind of thinking. For him, such thoughts were means toward an end. The end was to elicit the repressed latent thoughts to which they were ultimately connected and, in this way, to cure his patients. The associations themselves, as Kaja Silverman (2009: 44) notes, were for him ultimately irrelevant. But, following Silverman (2009: 65), there is another way to think about such chains of associations.[18] Rather than arbitrary, and pointing only inward toward the psyche, we might see these associations as thoughts in the world—exemplars of a kind of worldly thinking, undomesticated, for the moment, by a particular human mind and her particular ends.

This is what Luisa's thinking offers. It is a kind of creativity that comes in the form of listening (Silverman 2009: 62), and its logic is central to how an anthropology beyond the human can better attend to the world around us. If Amériga was forcing thought to yield its return, Luisa allowed the thoughts of the forest to resonate somewhat more freely as they moved through her. By keeping her imitation of the antbird's call below the symbolic level, holding its potential stabilized "meaning" in abeyance, Luisa allowed the sonic form of this vocalization to propagate. Via a chain of partial sonic isomorphisms, *"chíriqui'"* drew in its wake a series of ecological relations with the effect that the traces of the feline were carried across space and species lines through the dense thickets to that place where Luisa was harvesting fish poison the moment her dogs were attacked.

UPFRAMING

The possibilities inherent to this kind of play notwithstanding, access up to a type-level perspective—being able to recognize the cuckoo's call *shicuá'* or the anteater's hiss as instances of the omen *Shicúhua*—is empowering. And this formal hierarchical logic is what informed the Jaminaua shaman's quest for apprenticeship downriver. By going downriver, he was able to see the particular river from which he hailed as just one instance of a broader, more general pattern. Through this process of "upframing," he was now privy to the view from a higher-order emergent level (a sort of "type") that encompassed the individual rivers and their villages, which can here be understood as the lower-order component parts (the "tokens") of this system. These properties of a logical hierarchy, instantiated in an ecosystem, are what allowed this shaman to reposition himself within a sociopolitical hierarchy.

Not surprisingly, then, relations between humans and spirits, like those between humans and animals, are structured by the hierarchical properties inherent to semiosis. Here too there is a nested, increased ability to interpret as one moves up the hierarchy. Recall, from the previous chapter, that although the Runa can readily understand the meanings of dog vocalizations, dogs can understand human speech only if they are given hallucinogens. Similarly, although we humans need hallucinogens to understand the forest masters, these spirits can readily understand human speech; the Runa need only talk to them, as, in fact, they sometimes do in the forest. Just as animal utterances can be seen as tokens that require a further interpretive step to be seen as conforming to a type, the limited perceptions that humans have of the spirit realm also need to be appropriately translated into a more general idiom to be understood in their true light. The Runa in their everyday life see the game animals that they hunt in the forest as wild animals. But they know that this is not their true manifestation. Seen from the higher perspective of the spirit masters who own and protect these creatures, these animals are really domesticates. What the

Runa see as a gray-winged trumpeter, chachalaca, guan, or tinamou is really the spirit master's chicken. Here too there is a hierarchy that assumes certain logical semiotic properties. All these wild birds, as the Runa experience them in the forest, are token instantiations of a more general type—the Chicken— as interpreted at a higher level. And this something more—this higher emergent level—is also something less. All those forest birds share something in general in common with a chicken, but treating them solely as the chickens that in some real sense they also are erases their particular species-specific singularities.

One could also say that the spirit master's perception of the bird requires less interpretive effort. Following Peirce's (CP 2.278) insistence that the chain of semiotic interpretance always ends in iconism because it is only with iconism, as Deacon (1997: 76, 77) underscores, that the differences that would require further interpretation are no longer noticed (it is with iconism, that is, where mental effort ends), we could say that there is less interpretive effort required by the masters who see the forest birds just as they really are—as domestic chickens. We humans, by contrast, would have to smoke lots of "strong" tobacco, take hallucinogens, or dream particularly "well," as people in Ávila would say, to have the privilege of seeing the different kinds of wild game birds encountered in the forest as the chickens they really are.

INSIDE

Spirit masters need not exert the interpretive effort we humans require because, like rubber floating down the river, or the congeries of animals attracted to a fruiting tree, or a port city teeming with the upriver riches that collect there, they are already inside this emergent form. In fact, people in Ávila often refer to the reality of the spirit master realm as *ucuta* (inside), as opposed to the everyday human realm, which is *jahuaman* (on the surface). Because the spirit master realm is, by definition, always inside form, the animals are always abundant there, even though we humans aren't always able to see them. The woolly monkey troop we encountered while hunting one day that I, with my binoculars, diligently estimated as consisting at most of thirty individuals, Asencio, a veteran hunter and careful observer of the beings of the forest, described as numbering in the hundreds. And those animals that are not ever seen in the forests around Ávila, like squirrel monkeys, which are abundant at lower, warmer elevations, or white-lipped peccaries, which are no longer found locally, are nevertheless said to be present "inside" the domain of the masters of the forest. It's not that the animals aren't there; it's just that the masters don't allow us to see them. They don't allow us inside the form that holds them.

Animal abundance is not the only thing that is unchanging in the spirit world. The realm of the masters is also a kind of afterlife, Marcelino's paradise. And the Runa who go there never age and never die. Not long after the young woman out fishing found her in the forest, Rosa returned to the realm of the masters—this time forever. Ventura later told me that when his mother died they "just buried her skin" (see chapter 3). That is, they buried her weather-worn, time-ravaged, maggot-eaten habitus—a sort of clothing that, in the manner of jaguar canines and white clothing, conferred on her, her particular earthly elderly affects. In the realm of the masters, Ventura explained, Rosa will always be a nubile girl, like her granddaughters, her body now immune to the effects of history (figure 8).

That Rosa will never age in the realm of the masters is also the result of the peculiar properties of form. History as we commonly imagine it—as the effects of past events on the present—ceases to be the most relevant causal modality inside form.[19] Just as the causes responsible for riverine and floristic spatial patterns are in a sense irrelevant to the ways in which these can be linked by a highly patterned emergent socioeconomic system, and just as the words in a language can relate to each other in a way that is largely decoupled from the individual histories of their origins, so too in the realm of the masters the linearity of history is disrupted by form. Pre-Hispanic chiefdom hierar-chies, cities, bustling market towns, and early-twentieth-century estates, of course, have their own unique temporal contexts. But they now are all caught up in the same form, and, as such, the particular histories of how and when they came to be become, in a certain sense, irrelevant. Form then, for a moment, and in a sense, "freezes" time.[20] All these differently situated historically con-tingent configurations now participate "ahistorically" in a self-reinforcing pat-tern that people in Ávila attempt to harness to get game meat.

As a regularity that can potentially exceed ontological domains and tempo-ral instances this kind of form, then, creates an emergent "always already" realm. What I mean by this is that one outcome of certain kinds of systems that capture and maintain regularity—whether a socioeconomic one that har-nesses physical and biological regularities, an expanding language that incor-porates terms from other vernaculars, or even the historically layered realm of

FIGURE 8. "Granddaughters" preparing peach palm beer *(chunda asua)*. Photo by author.

the spirit masters of the forest—is that they create a domain of circular causality in which the things that have already happened have never not happened. Take the English language, for example. We know that any given sentence might include words of, say, Greek, or Latin, or French, or German origin, but these histories are irrelevant to the "timeless" way such words come to give each other meaning by virtue of the circular closure of the linguistic system of which they now form part. My point is that, like language, these other, not necessarily human and not necessarily symbolic systems that I have been discussing also create an emergent realm partially decoupled from the histories—the past's effects on the present—that gave rise to them.

The always already realm of the forest masters captures something of the quality of being inside form. According to people in Ávila, "the dead," when they go ucuta, or inside, the spirit realm of the masters, become "free." "*Huañugunaca luhuar*," they say; "The dead are free." *Luhuar*, which I'm glossing as "freedom," is derived from the Spanish *lugar*, whose primary meaning is "place." But *lugar* also has a temporal referent. The phrase *tener lugar*, although today infrequently used in Ecuadorian Spanish, means to have the time or opportunity to do something. In Quichua *luhuar* refers to a domain where worldly spatiotemporal constraints are relaxed. It is a sort of realm where cause-and-effect no longer directly applies. To become luhuar, as people in Ávila explain, is to become free from earthly "toil" and "suffering,"[21] free from God's judgment and punishment,[22] and free from the effects of time. Inside this perpetual always already realm of the masters in the forest, the dead just carry on—free.

Humans do not just impose form on the tropical forest; the forest proliferates it. One can think of coevolution as a reciprocal proliferation of regularities or habits among interacting species (see chapter 1).[23] The tropical forest amplifies form in myriad directions thanks to the ways in which its many kinds of selves interrelate. Over evolutionary time organisms come to represent with increasing specificity environments made ever more complex through the ways in which other organisms come to more exhaustively represent their surroundings. In neotropical forests this proliferation of habits has occurred to a degree unmatched by any other nonhuman system on this planet (see chapter 2). Any attempt at harnessing the living beings of the forest is wholly dependent on the ways in which such beings are embedded in these regularities.

As I said, this ubiquity of form does something to time. It freezes it. There is something, then, to Lévi-Strauss's much-maligned characterization of Amazonian societies as "cold"—that is, as resistant to historical change—in juxtaposition to those "hot" Western societies that supposedly embrace change (Lévi-Strauss 1966: 234).[24] Except what is "cold" here is not exactly a bounded society. For the forms that confer on Amazonian society this "cold" characteristic cross the many boundaries that exist both internal to and beyond human realms. The early-twentieth-century international rubber economy was just as constrained by the forest's forms as is Ávila hunting. Like kinds (see chapter 2), form need not stem from the structures we humans impose on the world. Such patterns can emerge in the world beyond the human. They are emergent with respect to the lower-order historical processes, those that involve the past's effects on the present, that give rise to them, and that also make them useful.

THE DETRITUS OF HISTORY

That the emergent forms of the forest are partially decoupled from the histories that gave rise to them does not banish history from the realm of the spirit masters of the forest. Bits of history, the detritus of prior formal alignments, get frozen inside the forest form and leave their residues there.[25] For example, *Tetrathylacium macrophyllum* (Flacourtiaceae) is a tree with a cascading panicle of translucent dark red fruits whose Quichua name, *hualca muyu*, means, appropriately, "necklace beads." However, rather than resemble the popular opaque glass necklace beads of Bohemian origin that have been a mainstay of Amazonian trade for the past century, these fruits bear a striking resemblance to an earlier dark red translucent Venetian trade bead that was in wide circulation throughout the colonial and neocolonial world. It passed through Ecuador around the time of the presidency of Ignacio de Veintemilla (1878–82) and is, accordingly, still referred to by some Ecuadorians as *veintemilla*. That the Ávila plant *hualca muyu* is linked to this nineteenth-century bead is the product of the peculiar time-freezing properties of form. The historical trace of a good that was traded and, like Simson's beads, commensurated with forest products remains caught in the always already form of the forest master's realm, even after people have long forgotten it. Another example: some kinds of demonic spirits, supai, that wander the forest are described as wearing priestly habits, even though today's local priests have long since abandoned wearing the black robe.

It is not, then, that history simply permeates the Amazonian landscape, as critical cultural geographers and historical ecologists contend as a counter to the romantic myth of a pristine wild Amazonian "nature."[26] Instead, the history that gets caught in the forest is mediated and mutated by a form that is not exactly reducible to human events or landscape.

The challenge for the Runa is how to access the forms of the forest that concentrate wealth. For in this always already realm animals exist in unchanging abundance. As with the Juruá-area shaman, the way they do this involves a process of upframing to see animals from the privileged (and objec-tifying) perspective of the masters—namely, not as singular selves, each with its own point of view, but as resources, and not as ephemeral subjects but as stabilized objects, owned and controlled by the master, a more powerful, emergent self. The Runa attempt to access the riches of the forest masters by mobilizing the disparate historical traces of strategies for negotiating with people more powerful than themselves that have gotten caught up—frozen, like Venetian trade beads and priestly habits—inside the master's form.

For example, it's been over a century and a half since the Runa had to pay regular tribute to government officials and clergy (Oberem 1980: 112), yet trib-ute still exists in the realm of the masters. When people kill a tapir they are required to offer trade beads as tribute to the spirit masters that own this ani-mal, so that these masters will continue to provide meat. Out on a hunting trip, Juanicu attempted to capitalize on the reciprocal obligations that this colonial arrangement entails. He offered the master tribute in the form of a few grains of corn tucked in the crevices of a tree base. When the master failed to provide us with game meat, as was his obligation, given that Juanicu had dutifully kept up his side of the bargain, Juanicu unashamedly reprimanded him—yelling, in the middle of the forest, "You're stingy!"—in exactly the same way I once heard him rebuke a politician visiting Ávila during election season who failed to give out cigarettes and drink.

On other occasions the Runa attempt to communicate with the masters through rhetorical formulas identical to those their sixteenth-century fore-bears used in negotiating peace contracts with the Spaniards. These include invoking a numerically parallel structure that attempts to make more balanced what, in another context, Lisa Rofel (1999) has called "uneven dialogues";[27] in the colonial case this involved making five demands in exchange for five con-cessions to the Spanish authorities, as is visible in one such late-sixteenth-century contract between the local indigenous chiefs and Spaniards (Ordóñez de Cevallos 1989 [1614]: 426). In the contemporary one, it is evident in the use of certain hunting and fishing charms that require a special ten-day fast—"five days for the master and five for the Runa," as people in Ávila put it.[28]

And the Quito-in-the-forest to which Rosa traveled is a reflection of more than four centuries of earnest attempts by Ávila-area people to negotiate with the powerful beings that live there for access to some of their wealth. Indeed, part of those sixteenth-century negotiations involved an attempt, unsuccessful alas, to convince the Spaniards to build a Quito in the Amazon—a request to which colonial documents (Ramírez Dávalos 1989 [1559]: 50, see also 39) and contemporary myths attest, and whose deferral continues to motivate desires to harness the riches that are harbored inside the forest.[29]

Each strategy for accessing the accumulated riches of the powerful has an independent causal history. But this no longer matters. They are all now part of something general, the forest master's form. And they can each serve as points of access to some of its riches.

It is not, however, just abundant game meat that such strategies promise. For they also hold out the possibility of some sort of access to the long and layered history of deferred desires that the quest for this meat has come to represent.

FORM'S EFFORTLESS EFFICACY

I hope here to have illustrated some of the peculiar properties of form, and I hope to have given some sense of why anthropology should pay form more attention. That it hasn't is, indeed, also an effect of form's peculiar properties. As anthropologists we are well equipped to analyze that which is different. However, as Annelise Riles (2000) has noted in her study of the circulation of bureaucratic forms associated with Fijian participants in a UN conference, we are less ready to study that which is invisible because we are "inside" it. Form largely lacks the palpable otherness—the secondness (see chapter 1)—of a traditional ethnographic object because it is only manifest qua form in the propagation of its self-similarity. "It is the people who are outside of the monastery who feel its atmosphere," writes the Zen master. "Those who are practicing actually do not feel anything" (Suzuki 2001: 78).

For these reasons it is much easier to understand the semiotic importance of indexicality—the noticing of difference—than it is to understand iconicity, which involves the propagation of regularities, through a specially constrained sort of indifference (see chapter 2). Perhaps this is why the propagation of similarity through indistinction is sometimes erroneously considered something other than representation. However, walking stick "twigginess" and a contagious yawn that propagates across bodies and even possibly across species lines (to give just two examples where iconicity is predominant) are semiotic phenomena, even though they largely lack an indexical component that can be interpreted as pointing to anything other than another instance of the patterns they instantiate. One could say that our habits become noticeable to us only when they are disrupted, when we fall outside of them (see chapter 1). And yet understanding the workings of that which is not noticed is crucial for an anthropology beyond the human. Form is precisely this sort of invisible phenomenon.

Form requires us to rethink what we mean by the "real." Generals—that is, habits, regularities, potential recurrences, and patterns—are real (see chapter 1). But it would be wrong to attribute to generals the kinds of qualities that we associate with the reality of existent objects. When I say that game birds are, from the perspective of the masters, really chickens, I am referring precisely to this way in which generals are real. The reality of the master's chicken is that of a general. And yet it has a possible eventual efficacy: it is able, as a sort of type, to index specific encounters with different kinds of birds, be they guans, chachalacas, or curassows. In this respect these encounters are not unlike the one I had with the peccary on that rainy day in the forest.

Without the day-to-day interactions that the Runa have with game birds, there would be no chickens in the realm of the masters. And yet the master realm enjoys a level of stability, which is partially decoupled from these day-to-day moments of forest interaction. This is why in the realm of the masters white-lipped peccaries can abound even though they haven't been found in the forests surrounding Ávila for many years now.

Although stable, form is fragile. It can emerge only under specific circumstances. I was reminded of this when I took a break from writing this chapter to prepare a pot of cream of wheat for my sons. Before my very eyes, the telltale self-organizing hexagonal structures known as Bénard cells, which form as liquid is heated from the bottom and cooled from the top under just the right conditions, spontaneously emerged across the surface of the simmering cereal. That these hexagonal structures promptly collapsed back into the sticky gruel is testament to form's fragility. Life is particularly adept at creating and sustaining those conditions that will encourage such fragile self-organizing processes to predictably take place (see Camazine 2001). This, in part, is why I have focused here on the ways in which complex multispecies associations cultivate form in ways that also think their ways through us when we become immersed in their "fleshliness."[30]

Form cannot be understood without paying attention to the kinds of continuities and connections that generals have with regard to existents. Accordingly, my concern here has been not just with form and those properties that make it unique—its invisibility, its effortless propagation, a kind of causality associated with it that appears to freeze history—but also with the ways in which form emerges from and relates to other phenomena in a manner that makes its unique properties come to matter in the worlds of living beings. I am not just interested in that which is "inside," but in how such an inside came to be, and also how it dissolves when the material conditions—be they riverbeds, parasites, or UN paychecks—necessary for its propagation cease to exist. And I am not just interested in form per se, but in how we "do things with" it. And yet doing things with form requires becoming infected with its causal logic, a logic that is quite different from that which is associated with the pushes and pulls of efficient causality, different, that is, from the ways in which the past affects the present. Doing things with form requires succumbing to its effortless efficacy.

None of this is to lose sight of the unique properties of form, and the possibility, as Riles notes, that anthropology might emerge from its crisis of representation by experimenting with ways of making the invisible "inside" more apparent. Building on Strathern (1995, 2004 [1991]), Riles's solution is to turn form "inside out." That is, she attempts to render form visible through an ethnographic methodology that amplifies it. Rather than try to make form apparent from an external perspective by indicating our discontinuities with it, she allows the patternings inherent to the proliferation of bureaucratic documents and the ones we academics might produce about them to multiply until their similarities become manifest.

I offer here no such aesthetic solution to the problem of elucidating form. I only wish to give a sense of some of the ways in which form moved through me. When I dreamed that night at Ventura's house of a peccary in a pen perhaps I too for a moment got caught up "inside" the forest master's form. What I would like to suggest is that the semiotics of dreaming, understood in terms of the peculiar properties of form I have explored here, involves the spontaneous, self-organizing apperception and propagation of iconic associations in ways that can dissolve some of the boundaries we usually recognize between insides and outsides.[31] That is, when the conscious, purposive daytime work of discerning difference is relaxed, when we no longer ask thought for a "return," we are left with self-similar iterations—the effortless manner in which likeness propagates through us. This is akin to Luisa's sonic web that linked the antbird to heliconias and to the jaguar that killed the dogs and all these to the humans in the forest whose dog that was—a web that emerged in the space of possibility that opened because she did not attempt to specify the meaning of the bird's call that she imitated. (Luisa was, in a sense, free.)

Considering it alongside this and the other various form propagations I've discussed here, I've come to wonder how much my dream was ever really my own; for a moment, perhaps, my thinking became one with how the forest thinks. Perhaps, like Lévi-Strauss's myths, there is indeed something about such dreams, which "think in men, unbeknownst to them."[32] Dreaming may well be, then, a sort of thought run wild—a human form of thinking that goes well beyond the human and therefore one that is central to an anthropology beyond the human. Dreaming is a sort of *"pensée sauvage"*: a form of thinking unfettered from its own intentions and therefore susceptible to the play of forms in which it has become immersed—which, in my case, and that of the Ávila Runa, is one that gets caught up and amplified in the multispecies, memory-laden wilderness of an Amazon forest.

SIX

The Living Future (and the Imponderable Weight of the Dead)

fire escapes old as you
–Tho you're not old now, that's left here with me
—Allen Ginsberg, *Kaddish for Naomi Ginsberg*

A tuft of fur snagged on a spine was the final clue that led us to the body of the peccary that Oswaldo shot several hours before. We were on Basaqui Urcu, a steep foothill of Sumaco Volcano northwest of Ávila. Swatting at the swarm of blood-sucking flies[1] inherited from our quarry we sat down to rest. As we caught our breath Oswaldo began to tell me what he had dreamed the night before. "I was visiting my compadre in Loreto," he said, referring to the market town and center of colonist expansion half a day's walk from Ávila, "when suddenly a menacing policeman appeared. His shirt was covered with clippings from a haircut." Frightened, Oswaldo awoke and whispered to his wife, "I've dreamed badly."

Fortunately he was wrong. As the events of the day would prove Oswaldo had in fact dreamed well. The hair on the policeman's shirt turned out to have augured killing the peccary whose body now lay beside us (after hauling a peccary carcass, bristles will cling to a hunter's shirt just like hair clippings). Nonetheless, Oswaldo's interpretive dilemma points to a profound ambivalence that permeates Runa life: men can see themselves as potent predators akin to powerful "whites" such as the policeman, yet also feel like the helpless prey of these same rapacious figures.

Was Oswaldo the policeman, or had he become prey? What happened that day on Basaqui Urcu speaks to the complexity of Oswaldo's position. Who is that frightening figure that is also so familiar? How can a policeman, a being so threatening and foreign, also be oneself? This uncanny juxtaposition reveals something important about Oswaldo's ongoing struggle to be and become, in relation to the many kinds of others he encounters in the forests around Ávila that make him who he is.[2]

These many kinds of others that "people" the forests around Ávila include the living ones that the Runa hunt and who on occasion hunt them. But their ranks are also filled by specters of a long pre-Hispanic, colonial, and republican history. These specters include the dead, certain demonic spirits (who might also prey on the Runa), and the masters of the animals—all of these continue in a different but nonetheless real way to walk those forests that Oswaldo traverses.

Who Oswaldo is cannot be disentangled from how he relates to these many kinds of beings. The shifting ecology of selves (see chapter 2) that he must constantly negotiate in his hunts in the forest, as well as on his visits to Loreto, is also inside him: it makes up his "ecology" of self.

More to the point, Oswaldo's dilemma speaks to the question of how to survive as a self and what such continuity might mean. How should Oswaldo avoid becoming prey, an *it*, dead meat, when the position of hunter—the *I* in this venatic relation—has now come to be occupied by outsiders more powerful than himself?

The Runa have long lived in a world where whites—Europeans and later Ecuadorian as well as Colombian and Peruvian nationals—have stood in a position of manifest dominance over them and where whites qua whites have been intent on imposing a worldview that justifies this position. Here is how a rubber-boom–era estate boss, living on the confluence of the Villano and Curaray Rivers, writes about another boss's attempts to make his Runa peons see things this way:

> In order to convince them of the superiority of the white man over the Indians by reason of our customs and knowledge, and to rid them of their hatred of the Spanish language, a neighbor of mine on this river, a rubber man, employer of many laborers, called together all the Indians one day and showed them a figure of Christ. "This is God," he said to them. Then he added: "Is it not true that he is a *viracocha* [white man] with a beautiful beard?" All the Indians admitted that he was a *viracocha*, adding that he was the *amo* [master] of everything. (Quoted in Porras 1979: 43)

The estate owner's take on Runa-white relations encapsulates a certain history of conquest and domination in the Upper Amazon that simply cannot be ignored. It is a historical fact that whites have come to be *los amos*—the masters—of "everything." In facing this colonial situation of domination as history, we might expect two responses. The Runa could simply acquiesce, accepting a subservient position. Or they could resist. However, as Oswaldo's dream already indicates, there is another way to live with this situation. And this other way challenges us to question our understanding of how the past shapes the present at the same time that it suggests a way of inhabiting a future.

Runa politics are not straightforward. Although domination is a historical fact, it is a fact caught up in a form (see chapter 5). As I explore in this chapter, it is caught up in a form that takes shape in the realm of the spirit masters of the forest—a realm whose particular configuration is sustained by the ways in which people like Oswaldo continue to engage the forest's ecology of selves in their own search for sustenance.

This realm of the spirit masters of the forest also sustains Oswaldo in a psychic sense. And there is no vantage from which he can escape or resist this condition. He is always already in some way or another "inside" its form. The political theorist Judith Butler alludes to such a dynamic in her observation that

> [t]o be dominated by a power external to oneself is a familiar and agonizing form power takes. To find, however, that what "one" is, one's very formation as a subject, is in some sense dependent on that power is quite another. We are used to thinking of power as what presses on the subject from the outside. . . . But if . . . we understand power as *forming* the subject as well as providing the very condition of its existence . . . then power is not simply what we oppose, but also, in a strong sense, what we depend on for our existence and what we harbor and preserve in the beings that we are. (Butler 1997: 1–2)

Butler contrasts the brutal aspect of power in its cold externality to the subtle but no less real ways in which power pervades, creates, and sustains our very being. For power, as Butler intimates, is not reducible to the sum total of brutal acts. Power takes on a general form even if it is also instantiated— palpably, painfully—in the world and on our bodies.[3]

This final chapter of *How Forests Think* seeks to ask, with attention to Oswaldo's predicament, what, following Butler, it might mean to be and become in "formation." But it reconfigures this question by reflecting on how our understanding of the ways in which power works itself through form changes when we recognize form as a kind of reality beyond the human.

In this regard, I build on my discussion of form from the previous chapter. Form is, as I argued there, neither necessarily human nor necessarily alive, even though it is captured and cultivated by life and even though form also proliferates in those dense ecologies of selves such as the ones that exist in the forests around Ávila. In chapter 5 I discussed how harnessing form involves being made over by form's strange mode of effortless efficacy—a kind of efficacy in which the past's effect on the present ceases to be the only causal modality at play. If we are made over in our harnessing of form's strange causal logic—the self that harnesses form does not just do so by pushing, pulling, or resisting—then what we mean by agency changes. And if agency becomes something different, then politics changes as well.

But to understand Oswaldo's predicament we need to think not just in terms of the logics of form that the forest amplifies but also in terms of form's relation to certain other logics intrinsic to life. For what ultimately is at stake for Oswaldo, as his dream makes manifest, is survival. And the problem of survival is one that concerns the living (for it is after all only the living who die). If form, as I discussed in the previous chapter, can sometimes have the effect of freezing time, in ways that change our understandings of causality and agency, life disrupts our commonplace understanding of the passage of time in a different way, and this too must be considered in trying to understand Oswaldo's predicament. For, in the realm of life, it is not just the past that affects the present, nor is time just frozen. Rather, life involves, in addition to these, the special ways in which the future comes to affect the present as well.

Let me illustrate this way in which the future affects the present in the realm of life with a simple example from the forest. In order for a jaguar to successfully pounce on an agouti she must be able to "re-present" where that agouti will be. This re-presentation amounts to an importation of the future— a "guess" at what the agouti's future position will be—into the present via the mediation of signs. Being semiotic creatures through and through (see chapter 2), "we" all always have one foot (or paw) in the future.

In this chapter I'm thinking about this intrinsic relationship that obtains between life and future by reference to what Peirce called a "living future" (CP 8.194). This living future, as I argue here, cannot be understood without further reflecting on the special links that life has to all the dead that make life possible. It is in this sense that the living forest is also one that is haunted. And this haunting gets, in part, at what I mean when I say that spirits are real.

Survival—how to go about inhabiting a future—this is Oswaldo's challenge. And the solutions he finds are inflected by the living-future logic that is amplified in the forests he traverses. But survival here for Oswaldo is also an all-too-human problem (see chapter 4), one in which questions of power cannot be avoided. And this makes the problem of survival also a political one; for it prompts us to think about how we can find other ways to harness the power that will ultimately sustain our being in a manner that enables "us" to grow and even to flourish.

This chapter, then, focuses on the realm of the spirit masters of the forest. It does so with particular attention to how that realm makes apparent some aspect of the ways in which life (human and nonhuman) is connected to death, continuity to finitude, future to past, absence to presence, supernatural to natural, and ethereal generality to palpable singularity. All of these, ultimately, say something about the formative connection a self has to its many others. My interest here is to see how these articulations, as they become expressed in the realm of the masters, amplify and render conceptually available to us some of the living-future logic of a thinking forest—a logic that can help us take anthropology beyond the human.

That Oswaldo at a certain moment in the forest can—perhaps must—be a white policeman, involves the particular and sometimes disjointed and even painful ways in which some aspect of his future self reaches back to affect him from this realm of the forest masters. In the process it exposes the logic of some of these articulations that I mentioned. This spirit realm that emerges from the life of the forest, as a product of a whole host of relations that cross species lines and temporal epochs, is, then, a zone of continuity and possibility: Oswaldo's survival depends on his ability to access it. And yet Oswaldo's survival also depends on the many kinds of dead and the many kinds of deaths that this spirit realm holds in its configuration and that make a living future possible. Who one might be is intimately related to all those who one is not; we are forever giving ourselves over and indebted to these many others who make us who "we" are (see Mauss 1990 [1950]).[4]

Although it emerges out of Runa histories of engaging with the many kinds of selves that people their world, the realm of the spirit masters is also something other than the product of these histories of engagement. This realm is a sort of afterlife, which is closely related to but not reducible to the life that has come before it. It is, in this sense, its own kind of emergent real—one that is neither natural nor exactly cultural.

I explore this emergent ethereal realm with specific attention to the ethnographic manifestation of some of its special properties as well as to the hopeful politics that it might harbor. My goal is to reflect more generally about what this realm beyond the living—one that emerges from the rich ecologies of selves that the forest houses—can tell us about the living logics that such a thinking forest reveals.

Venturing beyond the living, as I do here, is important for the anthropology beyond the human that I have been trying to develop, for it is with attention to this realm of the spirit masters of the forest that we can better understand what continuity might mean and how best to face that which threatens it. In short, attending to what those spirits of the forest can teach us about continuity, growth, and even "flourishing" can allow us to cultivate other ways of thinking about how "we" might find better ways to live in the living future.

ALWAYS ALREADY RUNA

A curious mural that adorned the walls of the multipurpose hall of the headquarters of FOIN, the federation that represents the Runa communities of Napo Province (figure 9), seems to describe a progression from Amazonian savagery to European civilization. At the far left of a lineup of five men stands a long-haired "savage" Indian holding a blowgun and what appears to be a shell horn of the sort used to call and mobilize kith and kin.[5] He is what we would consider "naked," though he wears a penis string, face paint, necklace, and arm, wrist, and head bands. The next man wears a loincloth, and the horn lies behind him on the ground; otherwise he looks nearly identical. Then stands a man who, in keeping with Runa fashion of the late nineteenth century, wears shorts and a small tunic or poncho. He has just a dab of face paint and tries to hide his blowgun behind his back. The next man in the progression is fully clothed. He wears shoes, long pants, and a crisp white short-sleeved shirt. He is handsome, and whereas the previous figures have tiny heads, no necks, and huge arms, this man's body is well proportioned. The blowgun that caused the previous man such shame now simply lies abandoned behind him. He is also the only one who offers any hint of a smile. This figure is the epitome of a contemporary Runa man in the imaginary of the labor-union–influenced FOIN leadership of the 1970s and 1980s, a leadership that came of age before the influx of the international NGOs, and one that had yet to become culturally or environmentally "conscious." He is a Runa campesino, neither ethnic nor elite, neither sylvan nor urban. The final figure emerging from this backdrop now littered with the discarded trappings of a timeless savagery wears glasses, a suit, and a tie. His hair is neatly parted down the middle, and he sports a pencil moustache—a carefully nurtured wisp of the facial hair that whites seem to have no problem producing in revolting but also awesome

FIGURE 9. "To make brutes into men, and men into Christians" (Figueroa 1986 [1661]: 249): This mural, which existed in the headquarters of the indigenous federation FOIN during the late 1980s, ambiguously illustrates the legacies of this colonial endeavor. Photo by author.

abundance. He has the slight build of someone who has spent too much time indoors. He stands at grim attention. He seems nervous. In his right hand he clutches a briefcase. Strapped on his left arm, a wristwatch inexorably marks off the minutes of a day that is inside a linear temporality of which this man is now very much a part.

In the late 1980s I did some volunteer work for the federation that for a time had me living in its headquarters. This mural covered one of the walls. One evening, to celebrate the end of a workshop, the participants, primarily Runa men and women from Tena and Archidona and the villages that surround these towns, which are much more urban and less oriented toward the forest than Ávila, held a party at the headquarters. The mural was the source of a running joke throughout the evening. Every so often someone, invariably male, would point to one of the "savage" Indians standing to the left of the handsome Runa man in the lineup to indicate the stage of drunkenness to which he had descended.

The mural speaks to the primitivist narrative that has guided both mission-aries and colonists in this region: before the arrival of Europeans naked "wild savages" were the Amazon's only inhabitants; through a process of "taming" that spanned the colonial and early republican periods and continues to this day some of these wild savages became civilized, clothed, monogamous, salt-eating, and unthreatening Runa; they became, according to the colonial termi-nology, *indios mansos,* or tame Indians (Taylor 1999). Survivals of what, according to this logic, would be the primordial wild substrate can still be found in certain isolated regions. Some members of the Huaorani ethnic group (sometimes still pejoratively referred to in Quichua as Auca), who are considered homicidal, polygamous, and naked, serve as the present-day mod-els for the depiction of savagery to the far left of the mural.[6] The seventeenth-century Jesuit priest Francisco de Figueroa succinctly described this colonial project of attempting to fashion a certain kind of person. The missionary goal, he wrote, is "to make" Amazonian "brutes into men, and men into Christians" (Figueroa 1986 [1661]: 249).[7] The revelers that night were playing with the inherited legacies of this attempt (see also Rogers 1995).

Many people in Ávila would not disagree with such distinctions between savage and civilized. They emphatically concur that being human in the right ways involves eating salt, wearing clothing, and refraining from homicide and polygamy (see also Muratorio 1987: 55). But they differ as to how—or even whether—to locate these traits in time. The missionaries saw the adoption of these traits as the result of a gradual process of "taming" a brutish Amazonian substrate. In Ávila, however, "civilized" attributes such as monogamy and eat-ing salt are primordial aspects of Runa humanity. The Runa have always already been civilized.

An Ávila diluvial myth illustrates this. When the great flood swept over the land many Runa managed to save themselves by climbing to the top of Yahuar Urcu, one of the highest peaks in the area. Other Runa attempted to escape by boarding canoes. The women on board twined their long hair in an attempt to moor themselves to the treetops still above water. When these lashings became undone the canoes floated downriver and came to rest in what is today Huaorani territory. There the clothing of those Runa eventually wore away, and they also ran out of salt. They began killing people and thus became the present-day Aucas. The Aucas, then, are not the primordial savages from which the Christianized Runa evolved. Rather, they are fallen Runa. They too were once salt-eating, clothed, and peaceful Christians. Although the Quichua term *Auca* is generally translated as "savage" or "infidel," it may be more accurate to think of Aucas as apostates. They are those who have abandoned their former Runa way of life.[8] The Runa have always already been Runa. "Savages," by contrast, became so as their canoes swept them down the flooded rivers, carrying them far away from their unchanging Runa homeland; they are the ones who fell out of form and into time.

The "Runa" man of the primitivist mural—made by his past, vanishing in the future—is not, then, exactly congruent with this other kind of being, this "always already" Runa of Ávila. What I am suggesting is that for the Ávila Runa the mural would not depict a progression leading elsewhere but an ongoing fugue around a central figure—a Runa self—who always already is what he will become even in his ongoing and open-ended becoming. This constantly changing self, who is also continuous with his past and potential future instantiations, points to something important about life, and flourishing, in an ecology of selves.

NAMES

We tend to think of a term like *the Runa* as an ethnonym, a proper noun used to name another. And this is how I've been using it throughout this book. For such a term to be deemed appropriate, standard anthropological practice dictates that it be the name the people in question use for themselves. This is why we do not refer to the Huaorani by their pejorative Quichua name "Auca." And "Runa," at least when modified by a place name, is certainly used as an ethnonym in Ávila to refer to Quichua-speaking inhabitants of Amazonian Ecuador. So, for example, "the San José Runa" refers to the people from San José de Payamino. And those from San José de Payamino call their Ávila neighbors "the Ávila Runa." Naming others is unavoidable.

And yet people in Ávila do not name themselves. They don't call themselves the Runa (or the Ávila Runa for that matter). Nor do they use the term *Kichwa*, the ethnonym currently employed in the contemporary regional and especially national indigenous political movement. If we treat "Runa" as a label—asking only whether it is the right label—something important is obscured; the Runa don't use labels for themselves. In a certain straightforward sense, Runa, in Quichua, simply means "person." But it does not merely function as a substantive to be co-opted as an ethnonym, a label.

Going back to the mural, the man beaming and wearing a crisp white shirt, standing between the "savages" and the "white man," is, by any account, "Runa."

From the primitivist point of view "Runa" here would be an ethnonym, a label for a waypoint in a historical process of transformation in which one kind of being is made into another, on the way to becoming still another. The Ávila take on this, however, would be different. The man in the crisp white shirt would still be "Runa," but the label would refer to something else, something less visible, less easily nameable than a cultural group from which one came. This man never became Runa; he has always already been Runa.

What I wish to suggest, and this is something I hope will become more evident as the chapter progresses, is that "Runa" more accurately marks a relational subject position in a cosmic ecology of selves in which all beings see themselves as persons. "Runa" here is the self, in continuity of form. All beings are, from their points of view, in a sense "Runa," because this is how they would experience themselves when "saying" *I*.

If we treat "Runa" as a substantive we miss the way it actually functions more like a personal pronoun. We usually think of pronouns as words that stand in the place of nouns. But Peirce suggests that we flip this relation. Pronouns are not substitutes for nouns; rather, "they indicate things in the directest possible way," by pointing to them. Nouns are indirectly related to their referents, and thus they ultimately rely on these sorts of pointing relations for their meaning. This leads Peirce to conclude that "a noun is an imperfect substitute for a pronoun," and not the other way around (1998b: 15). I want to suggest here that the Runa man who is the subject of the mural is—on the Ávila take—functioning as a special kind of first-person pronoun: an *I*, or perhaps more accurately an *us*, in all its coming possibility.

As a noun "Runa" is an "imperfect substitute for a pronoun." In its imperfection it carries the traces of all the others with whom it has become an *us* in relation. What it is, and what it might become, is shaped by virtue of all the predicates—eating salt, monogamy, and so on—it has acquired, even though it is also something other than the sum total of these.

An *I* is always in some sense invisible. By contrast, it is the other—the *he*, the *she*, the *it* objectified—that can be seen and named. I should note that the third person—the other—corresponds to Peirce's secondness. It is what is palpable, visible, and actual because it stands outside us (see chapter 1). This in part explains why self-naming is so rare in Amazonian ecologies of selves. As Viveiros de Castro has observed, naming is really reserved for others: "ethnonyms are names for third parties; they belong to the category of 'they' not the category of 'we'" (1998: 476). It is not a question, then, of which ethnonym to use but a question of whether any ethnonym can capture a self's point of view. Naming objectifies, and that is what one does to others—to *its*.[9] The Runa— I'm slipping back into using the objectifying label—are not the *its* of history. They are *Is*, part of an ongoing *us*, alive, in life, surviving—flourishing.

Runa-as-*I*, as-*us*, is not a thing, to be affected by the past in the cause-and-effect ways in which things are. The Runa are not the objects of history. They are not its products. They were not made by history in this cause-and-effect sense. And yet who they are is an outcome of a certain intimate relation to the past.

This relation involves another kind of absence as well. It involves a relation to the absent dead. In this regard, the Runa are like the cryptic Amazonian insect known as the walking stick, which comes to be increasingly invisible in its growing confusion with twigs by virtue of all those other beings that it is not. Those other, somewhat less "twiggy" walking sticks, are the ones who become visible and in their visibility become the tangible, actual objects—the others, the *its*—of predation in such a way that the potential future lineages of those who remain invisible can continue on, hidden and yet haunted (by virtue of this constitutive absence) by these others that are not them.

AMO

Oswaldo's continuity as an *I*, as Runa, requires that he be a puma—a predator. He must be the hunter and not the hunted peccary he feared he would become when he encountered that policeman covered in hair clippings standing at his friend's door. Puma, recall, is often hypostasized as jaguar—its primary exemplar—although it more accurately marks a relational position of a self, an *I* continuing as *I* and alive, thanks to an objectifying relationship to other selves that this self creates through predation. As such, like "Runa," it too functions as an "imperfect substitute for a pronoun." Oswaldo is—must be—runa puma, a human-jaguar, to persist.

In Ávila, *runa puma* is synonymous with a kind of maturity of self. Many men, and many women too, cultivate a sort of becoming-puma, so that, after death, after their human skins are buried, they enter a jaguar body, to continue on, as a self, and an *I*—an *I* that is invisible to themselves, yet able to see others as prey, while being seen by these others as predator. One cultivates this puma nature not only with regard to one's postmortem future but, perhaps more important, so that this future puma will also inform one's present ability to continue living as a self; becoming-puma is a form of worldly empowerment.

And yet predation is a fraught form of relating, not without its own anxieties. A few months after killing the pig, Oswaldo dreamed of another such encounter. In this encounter he didn't have his gun. All he had was an empty refillable shotgun cartridge. Somehow he managed to shoot his quarry by blowing through the little hole at the cartridge's base as if it were a blowgun.[10] To his dismay, he suddenly realized that the "prey" he had shot in this fashion was not a pig but a friend from Loreto. Wounded in the neck, this friend ran to the safety of his house, only to emerge shortly after, now armed, and in pursuit of Oswaldo. There is something unmanageable, chaotic, and amoral about predation. It is a kind of power that can come back to haunt you.

In the 1920s Runa from the Napo River told the explorer and ethnographer Marquis Robert de Wavrin of how, many generations ago, some shamans escaped Spanish domination by putting on jaguar skins—"black ones, spotted ones, yellow ones"—and in this way becoming puma. Having become predators and living deep in the forest, they managed to evade the Spaniards, but they also began to turn on their fellow Runa—first by hunting the unfortunate hunters who ventured out in the forest and then by attacking their own Runa villages (Wavrin 1927: 328–29).

It is not entirely clear why predation has come to be such an important means of relating in Amazonia. Certainly there are many other forms of trans-species relating; it was, for example, through a parasitic—not a predatory—relationship that Oswaldo's and my blood became commingled with each other's and with that of the peccary in the forest that day as the swarm of blood-sucking flies that had been living off of Oswaldo's prey sought out a new host. But predation obviously resonates with hunting as much as it does with a colonial past and the social hierarchies that are its product. Being a predator, having to be so, is a frightening prospect, not free of its own ambivalences.

If Oswaldo is to be a successful hunter, if he is to continue, it is not enough for him to be a predator; he must also be "white." That is, if whites are hunters, which is manifestly true, given their history of preying on the Runa—it is the whites who hunted down their forebears with dogs and enslaved them during the rubber boom—then Oswaldo must also occupy this position when he sees himself as an *I*. The only other choice would be to become an object. The Runa must always already be Runa, puma, as well as "white."

More than white, they must, to be more accurate, always already be masters, *amos. Amo* means "master," "lord," or "boss" in Spanish and has served traditionally as a term of address for estate owners and government officials. The power this title indexes is indelibly linked to whiteness. In the mid-nineteenth century, for example, a man of African descent named Goyo was appointed governor of the Amazonian administrative region (known at the time as Oriente Province). Because this new governor was black, the Runa refused to treat him as a master. He was therefore forced to ask the previous governor, Manuel Lazerda, to continue as acting governor. As Lazerda recounts:

> The Indians believe that blacks are damned, charred in the infernal fires. They'll never obey Goyo. I'm his friend and I'll do his bidding. The earnings [primarily from forced sales to the Indians] will be divided in two parts: one for me, and one for him. Alone he wouldn't be able to do a thing. The catechized Indians will never recognize him as their *apu*.
>
> —What does *apu* mean?
> —*Amo, señor*. I will be for them their real master and lord. (Avendaño 1985 [1861]: 152)

In Ávila today, *amo*—*amu* in Quichua—remains inextricably associated with whites, the "real" masters and lords. But *amu* has also come to mark another I's perspective as appreciated from an external vantage point. And like "Runa" and "puma," it functions as an "imperfect substitute for a pronoun." That is, *amu* functions as a pronoun, but in the process it pulls in its wake all the predicates associated with the colonial history of domination to which it is linked.

Here is how Narcisa employs the term in her reflections on an encounter, discussed in chapter 3, that she and her family had in the forest with some red brocket deer and the propitious dream that preceded it.

"cunanca huañuchichinga ranita," yanica amuca
"therefore, I'll be able to make him kill it," I—the *amu*—thought

Thanks to what, earlier in the conversation she described as her "good dreaming," she felt certain she would easily be able to get her husband to kill at least one of the deer they encountered. *Amu*, here combined with the topic-marking suffix *-ca*, highlights the fact that her dreaming (and not the actions of her husband, as her interlocutors might otherwise expect) was what was important.[11] Her husband, who was to shoot the deer, was simply a proximate extension of her agency. This is why she—the amu—is the topic of this phrase. *Amuca* encourages us to note the not entirely expected fact that we should understand the events of the forest that day as revolving around her agency. Her dreaming self (which her narrating self, from a somewhat external position, can regard as amu), and not her husband with his gun, is the locus of cause. It is no coincidence that a word whose original and continuing meaning is "white lord" is used to denote this fact.

Because all selves and not just human ones are *I*s, *amu* also marks the subjective viewpoints of animals. After Maxi described to Luis how he had fired at an agouti from his hunting blind, Luis asked him:

> *amuca api tucuscachu*
> and the *amu* [—that is, the agouti—], was he hit?

Maxi responded, "Yeah . . . right in the back bone." "*Tias*," interjected Luis, using a sound image (see chapter 1) that simulates lead shot deftly cutting through the unfortunate agouti's flesh and bone—"sliced right through."[12] *Amuca* in this exchange shifts the topic of discussion from a focus on Maxi's action to the fate of the agouti-as-*I*.

The term *amu*, referring to a title that the Runa, as Lazerda observed, would bestow only upon a white person, now also refers to any Runa *I*. But because all beings, and not just humans, see themselves as *I* (and therefore, in a sense, also as Runa) it follows that they also all see themselves as masters. Whiteness is now understood as inseparable from one's sense of self when "saying" *I*, even when the one "saying" *I* is not human.

Amu, like Runa and *puma*, marks a subject position. And all of these nouns, which we might otherwise only take to mark, respectively, white, indigenous, or animal essences additionally mark a vantage point—the position of the *I*. The term *amu*, without losing its historical association to particular people with particular physical characteristics and a particular position in a power hierarchy (in fact, because of these accumulated associations), has also come to mark any self's point of view. The living *I*, the self, any self—qua self—in this ecology of selves, is amu. That self is by definition a master, and therefore in a certain sense "white."

This particular "imperfect substitute for a pronoun" has unique qualities. Along with puma (or white), amu invokes hierarchy. But it does so in a way that catapults the self into a plane that goes beyond that of the living. And this fact has important implications for what it is to be an *I*, in continuity.

Like Oswaldo and his ambivalent relation to the policeman, the Runa both are and are obviously not "the masters of everything." *Amu* captures something of this disjointed and alienated nature of the self's relation to itself. The masters have always already been right there, along with the Runa, not just in the realm of the living but also in those realms that span beyond life. The spirits, who control the animals and who live in that timeless always already realm deep in the forest, are known by many names in Ávila, but mainly they are simply called "the masters"—*amu-guna*. These forest masters appear to the Runa in dreams and visions as white rubber estate bosses or Italian priests. It is from the master's vantage point—when the Runa manage to inhabit it— that they are able to hunt successfully. When Oswaldo comes to recognize that he is the white policeman of his dream, he is not just becoming one of those officers who walk the streets of the towns of, say, Tena or Coca; he is also becoming a master of the forest and, in the process, inhabiting, in some way or another, this realm of the spirits.

The Runa, as always already Runa, have always already been in such intimate relation with these sorts of figures that populate the timeless realm of the masters. In mythic times the masters were always already there, as a pair of Christian apostles, who function as "culture heroes" and who walked the earth and guided the Runa.[13] Being guided by master-apostles involves a degree of intimacy mixed with separation and alienation. According to one diluvial myth, recounted by the early-twentieth-century Napo-area Runa (Wavrin 1927: 329), in mythic times the Amazon was inhabited by God and the Saints. During the flood God built a steamboat, which he used to escape up to heaven along with these Saints. When the flood receded God's now-abandoned boat washed up in the land of the foreigners. By observing this boat, the foreigners learned how to make ships as well as other machines. The original owners of modern technology may be white deities, but they have also always already been Amazonian and an intimate if also detached aspect of Runa life.

Let me explain what I mean about this relationship between intimacy and detachment. That the Runa are amu when "saying" *I* (and that they also stand in an intimate yet detached and sometimes subservient relation to those amu who inhabit an always already realm) distributes the self and marks the pain of those disjunctures that separate its successive instantiations.

Regarding such successive instantiations of the self, linguistic anthropologists working with Gê and Tupi Guarani peoples of central Brazilian Amazonia have noted that the first-person singular—the *I*—used in certain narrative performances can sometimes refer to the skin-bound self performing the myth or song. Whereas at other times it can refer to other skin-bound selves through quotation, and at still other times to a self that is distributed over a lineage that includes both the performer and the performer's ancestors (Urban 1989; Graham 1995; Oakdale 2002; see also Turner 2007). Regarding the latter, Greg Urban (1989: 41) describes how a Shokleng origin myth-teller enters a trance-like or possessed state when embodying the *I* of his ancestors. Urban refers to this special kind of self-reference, in which the self is also a lineage, as a "projective *I*." It is projective because by embodying these "past *I*s" the narrator also comes to embody the "continuity" (45) of his self—a self that has now become part of a more general "emergent" lineage of selves (42).[14] His *I* becomes an *us*.

I want to suggest that *amu* captures something important about this "projective *I*." It refers to the self in continuity—an "*us*" with its "indefinite possibilities" (Peirce CP 5.402; see chapter 1). This continuity does not just stretch back to the ancestors. It also projects into the future. And it also captures something about how the *I* is constitutively related to a *not-I*—to the whites, the spirits, and the dead that the living Runa are but also are not.

BEING *IN FUTURO*

The Runa self is always already Runa, puma, and especially always already master, or amu. This self always has at least one paw in a spirit realm, which is neither located just in the present nor the simple product of the accretions of its cumulative pasts. There is a formal semiotic logic to this. As I argued in the first chapters of this book signs are alive and all selves, human and nonhuman, are semiotic. What a self is, in the most minimal sense, is a locus—however ephemeral—for sign interpretation. That is, it is a locus for the production of a novel sign (termed an "interpretant"; see chapter 1) that also stands in continuity with those signs that have come before it. Selves, human or nonhuman, simple or complex, are waypoints in a semiotic process. They are outcomes of semiosis as well as the starting points for new sign interpretation whose outcome will be a future self. Selves don't exist firmly in the present; they are "just coming into life in the flow of time" (Peirce CP 5.421) by virtue of their dependence on future loci of interpretance—future semiotic selves—that will come to interpret them.

All semiosis, then, creates future. This is something distinctive about self. Being a semiotic self—whether human or nonhuman—involves what Peirce calls "being *in futuro*" (CP 2.86). That is, in the realm of selves, as opposed to in the inanimate world, it is not just the past that comes to affect the present. The future, as, I discussed in this chapter's introduction, as it is re-presented, also comes to affect the present (CP 1.325; see also CP 6.127 and 6.70),[15] and this is central to what a self is. The future, and how it is brought into the present, is not reducible to the cause-and-effect dynamic by which the past affects the present. Signs, as "guesses," re-present a future possible, and through this mediation they bring the future to bear on the present. The future's influence on the present has its own kind of reality (see CP 8.330). And it is one that makes selves what they are as unique entities in the world.

Peirce refers to the past—the product of causes and effects—as fixed or "dead." Being in futuro, by contrast, is "living" and "plastic" (CP 8.330). All semiosis, as it grows and lives, creates future. This future is virtual, general, not necessarily existent, and yet real (CP 2.92). All selves partake of this "living future" (CP 8.194). Neotropical forests, such as those around Ávila, proliferate semiotic habits to a degree unprecedented in the biological world, and in the process they also proliferate futures. This is what humans—the Runa and others as well—step into when they enter the forest and begin to relate to its beings.

And yet the kind of future that humans create is emergent with respect to the sorts of futures that characterize the nonsymbolic semiotic world in which such a future is nested. Like an icon or an index, a symbol must come to be interpreted by a future sign potentially coming into being in order for it to function as a sign. However, a symbol additionally depends on these future signs for its very qualities: Its "character . . . can only be realized by the aid of its [i]nterpretant" (CP 2.92). For example, the phonological qualities of a word like *dog* are arbitrary and are only fixed by virtue of the conventional relation the word has to a vast virtual, ethereal, and yet real realm of other such words (and their contrastive phonological qualities) that provide the context for its apperception and interpretation (see CP 2.304; see also 2.292–93). By contrast icons and indices retain their qualities (but not their ability to function as signs) independent of their intepretants. An icon, such as the Quichua sound image "*tsupu*," would retain the sonic qualities that make it significant, even, without the existence of those entities that plunge—tsupu—into water or whether it is ever interpreted to sound like such plunging entities. Although the qualities that make an index significant depend on some sort of correlation with its object of reference, like an icon it would retain these characteristics even when it is not interpreted as a sign. A palm tree crashing down in the forest would still make a sound even when no one—not even a skittish woolly monkey—is around to take this crash to be an index of danger (see chapter 1).

In sum, unlike an icon or index, a symbol's very being qua symbol relies on the emergence of a whole host of not necessarily existent and yet real signs that will come to interpret it. It is doubly dependent on the future.

The realm of the masters amplifies this being in futuro logic, which is central to all of semiotic life, at the same time that it is also made into something else by human symbolic semiosis. For Oswaldo to remain a living sign, he must be able to be interpretable by this virtual, yet real, realm of the masters—a realm where he needs to be treated as an *I* and not an *it* to survive. He must, in short, be capable of being hailed by a master as a *you*. And this will only be possible when in the realm of the masters he too actually becomes an *I*, in futuro.

This virtual realm of the masters is physically located deep in the forest. It emerges out of the forest's living ecology of selves—an ecology that is itself creating proliferating networks of futures. These proliferating networks come to shape the future realm of the masters. And so this spirit realm comes to capture the logic of a "living future" in a way that cannot just be explained in terms of the language or culture of its human participants. And this makes this realm more than a symbolic gloss on a nonsymbolic nonhuman world.

Amu, I want to suggest, is a particular colonially inflected way of being a self in an ecology of selves filled with a growing array of future-making habits, many of which are not human. In the process, amu renders visible how a living future gives life some of its special properties and how this involves a dynamic that implicates (but is not reducible to) the past. In doing so, amu, and the spirit realm upon which it draws its power, amplifies something general about life—namely, life's quality of being in futuro. And it ratchets this quality up a notch; the spirit realm of the masters is "more" in futuro than life itself. The realm of the spirits amplifies and generalizes this living-future logic, and it brings it to bear on an everyday political and existential problem: survival.

AFTERLIFE

Regarding the view of the afterlife held by one eighteenth-century Upper Amazonian group known as the Peba, the Jesuit missionary priest Juan Magnin (1988 [1740]: 477) reported with exasperation, "Their take on the matter is unequivocal. They say . . . they are all Saints; and that none of them will go to Hell, instead they'll all go to heaven, where their relatives are, Saints like them." Missionaries had little trouble getting the forebears of the Runa and other Upper Amazonians such as the Peba to comprehend heaven. And yet, to their continuing chagrin, they found that the locals insisted on understanding this afterlife realm as unfolding in a forest of all-too-earthly plenty—one that, according to a bemused missionary working among the Runa, has "rivers that contain more fish than water" and, most important, "astronomical quantities" of manioc beer (Porras 1955: 153). Seventeenth- and eighteenth-century accounts resonate with contemporary ones: This "other life," where the Indians "never die" (Figueroa 1986 [1661]: 282), provides "manioc in great abundance, and meat and drink as much as they wish" (Magnin 1988 [1740]: 477).[16] It is one with "no lack of steel axes and trade beads, monkeys, drinking parties, flutes and drums" (Magnin 1988 [1740]: 490; see also Maroni 1988 [1738]: 173).

Hell is an altogether different matter. It has been a continuing source of concern for missionaries, from Father Magnin's time and even earlier, that many Upper Amazonians were unwilling to conceive of damnation in Hell as a form of personal punishment for worldly sins. For the Runa, as many reports over the years attest, there simply is no Hell.[17] Hell, according to them, is where others suffer, especially whites and blacks.[18]

After Ventura's mother, Rosa, died she went "inside" to the world of the spirit masters (see chapters 3 and 5). She married one of those lords and became one of them—an amu. Her old sagging body—sloughed off like a snake's skin—was all that she left behind for her children to bury. Ventura's mother had died quite old, but now, her son explained, she lives eternally young in the realm of the masters. "[F]ire escapes old as you," wrote Allen Ginsberg, in his irreverent prayer poem mourning his own mother, "–Tho you're not old now, that's left here with me." Ventura's mother too was not old now. Never to die again, and never to suffer, she had become again—and now forever—like her pubescent granddaughters.[19] All that was left with her son was her aged body, decrepit like a rusted fire escape.

By becoming a master, Rosa, in a sense, became a Saint. She went to live forever in that realm of eternal abundance, full of game and beer and worldly riches, in that Quito deep inside the forest. She would never to go to Hell, she would never again suffer, and she would be forever free. As I discussed in the previous chapter, Rosa entered inside a form—that always already realm of the masters—where the impacts of time, the past's effects on the present, become less relevant. But Rosa is not the only Saint: "we are all Saints," insisted the Peba Indians who so frustrated the eighteenth-century Jesuit missionary.

I want to unpack this suggestion that Rosa is a Saint, and I want to even explore the possibility that we selves might all be Saints. I do so by attending to the relation that selves like Rosa have to the emergent virtual and "in futuro" realm of the masters. This is a realm of future possibility in which what it is to be an *I*, a self, is also shaped by the many kinds of dead, their many kinds of bodies, and the histories of their many deaths. That Rosa really continues on as a master, and perhaps as a Saint, however, is not just the direct effect of these others. For her continuity only becomes possible by virtue of a negative relation to them. It is an outcome that is not directly affected by the palpable presences of all those others but by their constitutive absences. I hope this will become clearer in the section that follows.

THE IMPONDERABLE WEIGHT OF THE DEAD

One day Juanicu went out with his dogs to the forest to collect worms for fish bait when he was badly mauled by a giant anteater. He nearly died from his wounds. Giant anteaters, known to rear up on their hind legs and slash out with the large curved claws of their forefeet when threatened, are truly formidable creatures; even jaguars are said to fear them (see chapter 3). Juanicu alternated between blaming his misfortune on a rival shaman with whom he has had an ongoing feud and, more mundanely, on his dogs, who led him to the animal (they were supposed to have stayed at home). Juanicu never blamed himself, nor did anyone else. Juanicu-as-*I* can never do himself harm. Only others can.

A young Ávila man, of whom I was very fond, was killed on the Huataracu River. They pulled out his body from the bottom of a deep pool. His chest was ripped open. He died while fishing with dynamite. No one doubted that. There was much less agreement as to the ultimate, or even proximate, cause of his death. Some blamed sorcerers and the darts and anacondas they sometimes send when attacking their enemies. Others blamed those responsible for the circumstances that led him to fish with dynamite on that day: a demanding brother-in-law; the fellow who gave him the dynamite; or the folks who took him out to the river. All established culpability with one person or another. Of the half dozen or so different explanations I heard, none put blame on the young man who died.

Omens reveal a similar logic. If the *camarana pishcu*, a kind of antshrike[20] that eats insects flushed by moving army ant colonies, is found flying around a house, someone will die; for this is how a child circles around her house crying inconsolably when she discovers that her mother or father is dead. The "grave digger" wasp[21] is known as such because it buries the tarantulas and large spiders it paralyzes (see Hogue 1993:417), throwing up fresh piles of red earth in the process, as if digging a grave. As with the antshrike, finding one of these near home is an omen that a relative will die. People in Ávila call such signs (and there are many)[22] *tapia*, bad omens. I first thought of these as omens of death, but I soon realized that they refer to something more specific: it is not death that they foretell but the deaths of others. In fact, they never augur the death of the person who finds them.

These examples say something about the counterintuitive relation of the self to that which it is not. Death for the self is ineffable, for the self is simply a continuation of life. The self is a general (see chapter 1). It is the experience of the death of others by the living that is so hard to bear, because it is what is palpable. "The thread of life is a third," wrote Peirce, whereas "the fate that snips it" is "its second" (CP 1.337; see also chapter 1).

The omens of mourning I have been discussing speak to the pain associated with another becoming other—a second, a thing—another that is no longer an *I*, no longer a possible part of a becoming-*us*-in-relation, or at least not for the moment. For the living mourners, death marks a rupture: the dead become *shuc tunu* or *shican* (different, other). The myth of the man being eaten alive by the juri juri demon that I recounted in chapter 3 explores the terrifying prospect of coming to experience oneself as such an object—an experience we will never have when we become objects.

But souls do not simply die; they can continue in that virtual future realm that living (and its attendant deaths) creates. The traditional kaddish—as opposed to Ginsberg's irreverent version—the Jewish prayer recited in memory of the dead, never mentions death.[23] Death can only be experienced from outside. Only others can snip at the thread of life. And only others, for the Runa, other kinds of people, especially blacks and whites (in the essentialist sense), go to Hell.

The self is always partially invisible to oneself in the sense that visibility requires objectification—secondness—and secondness misses something crucial about what a living self is. The *I* is an *I* because it is in form—because it partakes in a general mode of being that exceeds any particular instantiation of itself. That Rosa will become a master (and a Saint) is what makes her a living self. An anthropology that focuses on difference—one that focuses on the "nots" and the "seconds" (see chapter 2)—cannot attend to this invisible continuity of the self.

In a similar fashion, although it is true that walking sticks are invisible thanks to a specific relation they have to all of their more visible and less twiggy relatives that were noticed, just focusing on those objectified others misses the continuing persistence of the invisible *I* in a form that, in hindsight, leaves us with a visible proliferation of something general that, in this case we can call "twigginess."

All signs involve a relationship to something not present. Icons do this in a way that is fundamental to their being. Recall from previous chapters that, although we generally think of it in terms of likeness, iconicity is really the product of what is not noticed. (For example, that we don't at first notice the difference between a walking stick and a twig.) Indices, by contrast, point to changes in present circumstances—that there is something other to which we must attend (another kind of absence). Symbols incorporate these features but in a special way: they represent via their relation to an absent system of other such symbols that make them meaningful.

Life, being intrinsically semiotic, has a related association to absence. What a living organism-in-lineage, in-continuity-of-*I*—to use the Amazonian concept—is, is the product of what it is not. It is intimately related to the many absent lineages that did not survive, which were selected out to reveal the forms that fit the world around them. In a sense, the living, like the walking stick we mistake for a twig, are the ones that were not noticed. They are the ones that continue to potentially persist in form and out of time thanks to their relationship to what they are not. Note the logical shift: the focus is on what is not present: the imponderable "weight" (I think the oxymoron captures something of the counterintuitive nature of this claim) of the dead.

All of life, then, houses, by virtue of these constitutive absences, the traces of all that has come before it—the traces of that which it is not. The invisible realm of the masters makes, to follow the counterintuitive logic, all of this visible. It is in the realm of the masters that the traces of those who have lived (the pre-Hispanic chiefs, the black-robed priests, the grandparents and parents) and that which has happened (the great sixteenth-century uprising against the Spaniards, the circulation of the old trade beads, the forced tribute payments) continue. And this is the future realm, the realm that gives interpretability to the (human) living one as well. The realm of the masters houses

all of the specters of the past. And it is in this realm that the timeless *I* continues, by virtue of its intimate relation to these absences.

The *I* is in form and outside of history (see chapter 5). This is why nothing
can happen to it. Heaven is a continuation of form. Hell is history; it is what
happens to others. Heaven is a realm where people are not subject to time.
They never age. They never die there. Only *its* can be in time. Only they can be
affected, subject to dyadic cause-and-effect, out of form, subject to history—
punished.

THE *YOU* OF THE SELF

The realm of the masters is the product of the many futures created by the
forest. But it is more than this. A word depends for its meaning on the emergence of a vast symbolic system that will come to interpret it. Something like
this is happening in the forest as well. The realm of the masters is that vast
virtual system that emerges as humans—in their distinctly human ways—
attempt to engage with the other-than-human semiosis of the forest. The
realm of the masters, then, is like a language. Except it is more "fleshly" (Haraway 2003) than a language—being, as it is, caught up in vaster swaths of nonhuman semiosis. It is also at the same time more ethereal. It is a realm that
is in the forest but also beyond nature and the human. It is, in a word,
"supernatural."

This spirit realm of masters comes to interpret, and thus permits and con-
strains, who and how an *I* can be, at the same time that it provides the vessel
for the continuity—the survival—of that *I*. In Ávila, whiteness has come to
mark this *I* point of view. It marks a relative position within a hierarchy that
spans the cosmos—a hierarchy that ranges from the nonhuman to the human
realm and from the human one to the realm of the spirits. Therein lies Oswaldo's predicament. On the one hand, the Runa have always already been white.
On the other hand, they recognize a variety of beings—policemen, priests,
and landowners, as well as animal masters and demons—whose superior position in a historically inflected cosmic hierarchy is indexed by their whiteness.

This realm of the masters, however, is not just about the *I*. "Between the
reflexive '*I*' of culture," writes Viveiros de Castro (and by "culture," I take him to
mean the vantage from which a self sees herself as such, sees herself, that is,
as a person), "and the impersonal '*it*' of nature, there is a position missing, the
'*you*', the second person, or the other taken as other subject, whose point of

view is the latent echo of that of the 'I'" (1998: 483). This *you*, for Viveiros de Castro, gets at something important about the supernatural realm—a realm, I would add, that is not just reducible to nature, nor is it one that is reducible to culture. It is a realm that, according to a formal hierarchical logic, is situated "above" the human realm it makes possible.

"Supernature," continues Viveiros de Castro, "is the form of the Other as Subject" (1998: 483). I would say that it is the place where one can be called into being by this higher-order other self that is both strange and familiar. This is the realm from which Oswaldo's policeman hailed him. It is also the realm where all selves can experience themselves as masters—amu. So when the term *amu* is used in Ávila, whether in self-reference as in Narcisa's case or to refer to a being, human or nonhuman, that is properly other, it is done precisely to invoke this other *I*, taken as other subject—one whose voice, however faint, is a "latent echo" of the *I* in futuro.

The challenge is to avoid becoming an object in the process of this interpellation. And this is a real danger. Fear of this is what led Oswaldo to initially conclude that he had dreamed badly when he dreamed of seeing the policeman greet him with hair clippings on his shoulders. It is also why one cannot, for example, look at a *huaturitu supai*, the bird-clawed demon garbed in priestly robes that wanders the forests clutching a Bible. For becoming a *you* of that *I* would permanently transport you out of the realm of the living (Taylor 1993; Viveiros de Castro 1998: 483). And yet a self that is not destabilized by the *its* and *yous* that it constantly confronts, a self that does not grow to incorporate these into a larger *us*, is not a living *I* but a dead shell of one.

The question for the Runa, then, is how to create the conditions that will assure that they can continue to inhabit an *I* point of view. How, that is, to get into this higher-order *you* that both is but is never fully one's *I*? The techniques they use to do this are shamanic. Such techniques extend a paw into the future in order to bring some of that future back to the realm of the living.

I want to emphasize that the historical condition of possibility for shamanism is the very hierarchy it attempts to tap. Without the colonially inflected predatory hierarchy that structures the ecology of selves, there is no higher position one can enter from which to frame one's own. Emblematic of how shamanism relates to the history of hierarchies in which it is immersed is the term *miricu*, one of the names for "shaman" in Ávila.[24] The power of this term resides in the fact that it is a bilingual pun. As such, it captures two concepts in two different registers simultaneously; it is a Quichua-ization of the

Spanish word for doctor (*médico*) and it contains the Quichua verb "to see" (*ricuna*), in its agentive form; *ricu* is a seer. Shamans can see like doctors, those vanguards of modernity armed with all the powerful weapons of medical science. But this does not necessarily imply a desire to become like a Western doctor. Shamanistic seeing changes what it means to see.

How does one inhabit the *you* perspective? How does one make it one's own *I*? One does so by donning what we might call clothing—the equipment, bodily accouterments, and attributes that allow a particular kind of being to inhabit a particular kind of world. Such equipment includes the canines and pelts of the jaguar (see Wavrin 1927: 328), the pants of the white man (see also Vilaça 2007, 2010),[25] the robes of the priest, and the face paints of the "Auca." And such clothing can also be shed. Rosa sloughed off her aged body when she died. And it is reported in Ávila that some men, encountering jaguars in the forest and unable to scare them off, have undressed themselves to battle them. In this way, the jaguar is forced to recognize that his power comes from his clothing and that underneath this he is a person.[26] This is why jaguars, as Amériga fantasized with vengeful glee after her dogs were killed by one of them, so fear the sound of machetes slicing *"tlin tilin"* through the vegetation of the forest. For this reminds the jaguars of just how effortless it would be for people to slice through their *cushma*, or tunic,[27] which is the kind of clothing jaguars take their hides to be.[28]

Another set of examples of shamanic equipment. At a wedding, a man from a nearby Runa community approached me and, without a word, began to rub his smooth cheek against my beard stubble. Soon after, another young man approached and asked me to impart some of my "shamanic knowledge" by blowing on the crown of his head.[29] On a number of occasions when we were sitting around drinking beer older men would suddenly put on my backpack and strut around and then ask me to take a picture of them carrying my pack as well as other kinds of equipment: a shotgun, an ax, a pail of manioc beer. And one man asked me to take a portrait of his family, with everyone dressed in their best clothing, and he, wearing my backpack.[30] These are all little shamanic acts—attempts to appropriate something of what is imagined as a more powerful *you*.

I want to make clear here that it is not that the Runa want to become white in any sort of acculturative sense. For this is not a matter of acquiring a culture. Nor is the whiteness of whites intrinsically fixed. This is not about race. The Spaniard Jiménez de la Espada learned this on a visit he made in the 1860s to

the Runa of San José de Mote, a now-abandoned village located in the foothills of Sumaco Volcano about a day's walk from Ávila.

> The women, despite my generosity in distributing crosses, medallions, and beads,
> when I jokingly told them that I would like to marry one of them, they replied that
> who would want that, since I was not Christian.... I was a devil. (Jiménez de la
> Espada 1928: 473)

Although the Runa depend on various kinds of white equipment in order
to be and to continue as persons, they do not always extend such personhood
to the actual whites they encounter. White is a relational category, not an
essentialist one. The jaguar doesn't always have the canines, and the whites
aren't always the masters.

THE LIVING FUTURE

That Oswaldo managed to kill the peccary instantiated—brought into
existence—a heretofore only virtual real, which made that act possible.
Oswaldo became the policeman that day in the forest, and in the process he
brought back something of that future realm—ambiguously adumbrated in
his dream—into the world of the present. The realm of the masters is real. It
is real because it can come to inform existence, and it is real as a general possibility not reducible to that which will have happened. Reality is more than
that which exists. The realm of the masters is something more than human
and cultural, and yet it emerges from a specifically human way of engaging and
relating to a living world that lies in part beyond the human.

Spirits are real (see also Chakrabarty 2000; de la Cadena 2010; Singh 2012).
How we treat this reality is as important as recognizing it as such; otherwise
we risk taking spirits to be a kind of real—the kind that is socially or culturally
constructed—that is "all too human" and all too familiar. I concur that gods
emerge with human practice (Chakrabarty 1997: 78), but that does not make
them reducible to or circumscribed by the human contexts in which such
practices unfold.

The spirit realm of the masters of the forest has its own kind of general
reality: it is the emergent product of the relation it has to life's living future
and it "ratchets up" some of the properties that life harbors. Properties like
generality itself, constitutive absence, continuity across disjuncture, and a
disruption of cause-and-effect temporal dynamics become so amplified in

the realm of the masters that they become, in a sense, visible even in their invisibility.

Appreciating how spirits are their own kind of real is important for an anthropology that will be capable of attending to the human in relation to that which lies beyond it. But to do so one must be willing to say something general about what makes spirits real—something that includes but also goes beyond the fact that other people take them to be real, that we should take that fact seriously, and that we should even be open to how these kinds of reals might affect us (see, e.g., Nadasdy 2007).

In treating the realm of the masters housed deep in the forests around Ávila as an emergent real, my wish is to rediscover the world's enchantment. The world is animate, whether or not we are animists. It is filled with selves—I daresay souls—human and otherwise. And it is not just located in the here and now, or in the past, but in a being in futuro—a potential living future. A specific comingling of human and nonhuman souls creates this enchanted realm of the spirit masters in the forests around Ávila—a realm that is reducible neither to the forest nor to the cultures and histories of those humans who relate to it, even though it does emerge from these and cannot persist without them.

Living selves create future. Human living selves create even more future. The realm of the masters is the emergent product of a human way of living in a world beyond the human. It is the product of so much interspecies relating, coming together as so often it does, in the hunt. It houses all that future-making in a way that is general, invisible, and haunted by all the dead. It is, perhaps, the future's future.

In that future—that super-nature—lies the possibility for a living future. In killing that pig and not being killed, Oswaldo survived. To survive is to live beyond life: super + vivre. But one survives not only in relationship to life but in relation to its many absences as well. "To survive," according to the Oxford English Dictionary, means, "To continue to live after the death of another, or after the end or cessation of some thing or condition or the occurrence of some event (expressed or implied)." Life grows in relationship to that which it is not.[31]

The fractured and yet necessary relationship between the mundane present and the general future plays out in specific and painful ways in what Lisa Stevenson (2012; see Butler 1997) might call the "psychic life" of the Runa self, immersed and informed, as it is, by the colonially inflected ecology of selves in which it lives. The Runa are both of and alienated from the realm of the spirit world, and survival requires cultivating ways to allow something of one's future self—living tenuously in the realm of the forest masters—to look back on that more mundane part of oneself who might then hopefully respond. This ethereal realm of continuity and possibility is the emergent product of a whole host of trans-species and transhistorical relations. It is the product of the imponderable weight of the many dead that make a living future possible.

Oswaldo's challenge of surviving as an *I*, as it was revealed in his dream and as it plays out in this ecology of selves, depends on how he is hailed by others. These others may be human or nonhuman, fleshly or virtual; they all in some way make Oswaldo who he is. Oswaldo's survival—like Rosa's ongoing presence in that Quito deep in the forest—speaks to the puzzles of life that the forest amplifies; it speaks to the continual emergence of lineage out of the configuration of the individuals that instantiate them (see chapter 5). And it speaks to the creation of a form that stands in constitutive absence to that which it is not.

The soul, nonspecific and yet real, lives in such a continuity of form (see Peirce CP 7.591; see also chapter 3). The soul is general. Bodies (situated, equipped, erring, animal—not here to be confused with animate) individuate (see Descola 2005: 184–85, citing Durkheim). This gets at something about living futures. For life, in some way or another, is always about this sort of continuity across disjuncture that souls exemplify.

And what of this particular future's future? That which plays out in the neotropical forests around Ávila? What of the future of a future whose instantiation and continuing possibility is premised on killing some of those beings that a dense ecology of selves harbors? The emergence of the spirit realm of the masters of the forest is the product of the relationships among the many kinds of selves that make up this thinking forest. Some of these relationships are filial, others rhizomatic; some are vertical, others lateral; some are arborescent, others reticulate; some are parasitic, others predatory; and, finally, some are with strangers, and others, with those that are intimately familiar.

This vast but fragile realm of relating, played out in the forest and in that future realm that houses the forest's many pasts, is a world of possibility as long as not too many of these relations are killed. Killing, as Haraway (2008) points out, is not the same as killing a relation. And killing may actually permit a kind of relationship. Once the killing ends a larger, much more lasting silence may well follow. The Runa have an intimate relationship with the forest and with a kind of animacy that enchants the world because they kill—because they are part of this vast ecology of selves in this way. And killing and killing relationship are two different things, just as are individual and kind, token and type, life and afterlife. In all of these instances the first is something specific, the second general; all of these are real. It is by attentively engaging with the many kinds of real others that people this thinking forest—the animals, the dead, the spirits—that this anthropology beyond the human can learn to think about a living future in relation to the deaths that make that future possible.

Beyond

Animals came from over the horizon. They belonged there and here. Likewise they were mortal and immortal. An animal's blood flowed like human blood, but its species was undying and each lion was Lion, each ox was Ox.

—John Berger, *Why Look at Animals?*

Beyond the horizon there lies a Lion, a Lion more Lion than any mere lion. And beyond saying "lion," which calls forth that Lion, lies yet another, who might just look back. And beyond this eyeing one, lies an undying one, one we call "Lion" because she is a kind.

Why ask anthropology to look beyond the human? And why look to animals to do so? Looking at animals, who look back at us, and who look with us, and who are also, ultimately, part of us, even though their lives extend well beyond us, can tell us something. It can tell us about how that which lies "beyond" the human also sustains us and makes us the beings we are and those we might become.

Something of the living lion can persist beyond its individual death in a lineage of Lion to which it also contributes. And this reality lies beyond a related one that it sustains: when we speak the word *lion* it contributes to, at the same time that it draws on, a general concept—Lion—to invoke a living lion. So, beyond the uttered "lion" (technically a "token") lies the concept (the "type") Lion; and beyond that concept lies a living lion; and beyond any such individual lion lies a kind (or species or lineage)—a Lion—that both emerges from and sustains the many lives of these many lions.

I want to reflect on this idea of a beyond and how it figures in an anthropology beyond the human. I opened this book with an Amazonian Sphinx, a puma, who also looks back and who thus forces us to think about how to account anthropologically for the reality of a kind of regard that extends beyond human ways of looking. This led me to rethink the riddle that antiquity's

Sphinx posed to Oedipus: What goes on four in the morning, on two at noon, and on three in the evening? And I approached this riddle with a question of my own: What difference does it make that the Sphinx's question is posed from somewhere (slightly) beyond the human? *How Forests Think* investigates ethnographically why it matters to see things from the Sphinx's point of view.

That Sphinx beckons us to think with images. And this, ultimately, is what *How Forests Think* is about: learning to think with images. The Sphinx's question is an image, a likeness of its answer, one that is thus a kind of icon. The riddle is like a mathematical equation. Consider something as simple as $2 + 2 + 2 = 6$. Because the terms on either side of the equals sign are iconic of each other, learning to see "six" as three "twos" tells us something new about the number 6 (see Peirce CP 2.274–302).

We can learn something by examining the way the Sphinx's question, as icon, impels us to notice new things about Oedipus's answer, the "human." Her question can draw our attention to the animality we share with other living beings (our four-legged legacy) despite our all-too-human symbolic (and hence moral, linguistic, and sociocultural) ways of being in the world (captured in the image of our two-legged human gait). And it can help us notice what the kind of life that extends beyond the human ("four in the morning") and the kind that is all-too-human ("two at noon") share in common: that "three-legged" elder-and-his-cane (whom we might learn to appreciate as "mortal and immortal," self-and-object) invokes three key attributes we share with other living beings. These are finitude, semiotic mediation (the "cane" we living beings all use as we feel our ways through our finite lives), and—I can now add—a peculiar sort of "thirdness" unique to life. This kind of thirdness is the general quality of being in futuro, which captures the logic of life's continuity and how this continuity is made possible thanks to the room each of our individual deaths can make for the lives of others. The image of hobbling off "over the horizon" houses this "living future" as well.

Thinking with images, as I do here with the Sphinx's riddle, and as I do throughout this book, with all kinds of images—be they oneiric, aural, anecdotal, mythic, or even photographic (there are other stories being "told" here without words)—and learning to attend to the ways in which these images amplify, and thus render apparent, something about the human via that which lies beyond the human, is, as I've been arguing, also a way of opening ourselves to the distinctive iconic logics of how the forest's thoughts might think their ways through us. *How Forests Think* aims to think like forests: in images.

Turning our attention to the Sphinx, making her, not Oedipus, the protagonist in our story, asks us to look anthropologically beyond the human. This is no easy task. Chapter 1, "The Open Whole," charted an approach for doing so by finding a way to recognize semiosis as something that extends beyond the symbolic (that distinctively human semiotic modality that makes language, culture, and society, as we know them, possible). Learning to see the symbolic as just one kind of representational modality within the broader semiotic field within which it is nested, allows us to appreciate the fact that we live in sociocultural worlds—"complex wholes"—that, despite their holism, are also "open" to that which lies beyond them.

But recognizing such an open only impels us to ask: What is this world beyond us and the sociocultural worlds we construct? And so the second part of the first chapter turned to a reflection on how we might think about reality as something that extends beyond the two kinds of real that our dualistic metaphysics provides us: our distinctively human socioculturally constructed realities, on the one hand, and the objective "stuff" that exists out there beyond us, on the other.

It is no coincidence that I speak here with my hands to describe the choices this dualistic metaphysics affords. For this dualism is as deeply ingrained in what it means to be human as is our human tendency to think in terms of the right and left hands (see Hertz 2007). And it is no coincidence that I placed the realm of society and culture on the first, and hence the right, hand and relegated the realm of things to the second hand—the hand we consider to be the weaker, illegitimate, and sinister (from the Latin for "left") one. For it is that which we take to be human (our souls, our minds, or our cultures) that currently dominates our dualistic thinking. And this consigns the realm of the others, the nonhumans (evacuated of animacy, agency, or enchantment) to the left hand (a hand that, nonetheless, has its own subversive possibilities; see Hertz 2007; Ochoa 2007).

This dualism is not just a sociocultural product of a particular time or place; it goes "hand in hand" with being human, given that our propensity for dualism (our "twoness," in the Sphinx's terms) is the product of the distinctive properties of human symbolic thought and the ways in which the logic inherent to that kind of thinking creates systems of signs that can come to seem radically separate from their worldly referents.

Thinking in twos, then, is ingrained in what it means to be human, and moving beyond this kind of handedness requires a real feat of defamiliarizing the human. That is, it requires us to undertake an arduous process of decolonizing our thinking. It asks us to "provincialize" language in order to make room for another kind of thought—a kind of thought that is more capacious, one that holds and sustains the human. This other kind of thinking is the one that forests do, the kind of thinking that thinks its way through the lives of people, like the Runa (and others), who engage intimately with the forest's living beings in ways that amplify life's distinctive logics.

Those living beings enchant and animate the forest. My claim about the reality of enchantment and animism beyond the human, and my attempt to flesh it out and mobilize it conceptually in an anthropological approach that can take us beyond the human, is my left-handed offering to counter what we take to be the "right" ways to think the human.

Chapter 2, "The Living Thought," sought to unpack the claim that lives, and hence forests, think. That is, it looked to forms of representation—forms of thought—beyond language, with specific attention to the domain beyond the human in which these exist. When we focus only on the ways in which distinctively human thoughts relate symbolically—which informs linguistic, cultural, and social relationality and how we think about it—we miss something of the broader associational logic of "living thoughts." That nonhuman living beings are constitutively semiotic makes them selves. These nonhuman selves think, and their thinking is a form of association that also creates relations among selves. Attending to this other form of thought as a kind of relation, feeling it even, at times, emerge as its own conceptual object, and opening ourselves to its strange properties (such as the generative possibilities inherent to confusion or in-distinction), propels us to imagine an anthropology that can go beyond difference as its atomic relational component.

"The Living Thought," then, established why it is so important that anthropology look beyond the human toward life. In chapter 3, "Soul Blindness," I began to observe how the death beyond life is also central to life. My focus here was on how death becomes a problem—a "difficulty of reality"—intrinsic to life, and how the Runa struggle to find ways to come to terms with this.

"Trans-Species Pidgins" is a pivotal chapter. Having ventured beyond the human, and without losing sight of what that offers, I steered this anthropology back to the "all too human"—clarifying why this approach that I advocate is an anthropological approach, and not, say, an ecological one that agnostically charts multispecies relations. In the Runa's journeys beyond the human, in their struggles to communicate with those animals and spirits that "people" that vast ecology of selves that extends beyond them, they don't want to stop being human. Accordingly, this chapter traced ethnographically the kinds of strategies necessary to move beyond human modes of communication in ways that also secure a space for a distinctively human way of being.

Central to our distinctive ways of being human (which result from our propensity to think through symbols) is that we humans, as opposed to other kinds of living beings, are moral creatures. This is something that is not lost on the Runa as they struggle to get by in an ecology of selves that is everywhere shot through with the legacies of an all-too-human colonial history. Put simply, we cannot afford to ignore this all-too-human realm as we move beyond the human. That said, learning to attend to the kinds of lives that exist beyond the human (and beyond the moral), in ways that allow the logics of life beyond the human to work their ways through us, is itself an ethical practice.

In its attempt to relate the all too human to that which lies beyond the human, "Trans-Species Pidgins" also reveals something about the concept "beyond" as an analytic. "Beyond," as I deploy it, exceeds, at the same time that it is continuous with, its subject matter; an anthropology beyond the human is still about the human, even though and precisely because it looks to that which lies beyond it—a "beyond" that also sustains the human.

If much of this book has been about moving beyond the human to the realm of life, chapter 5, "Form's Effortless Efficacy," sought to move beyond the realm of life to the strange workings of form that sustain both human and nonhuman life. This chapter, then, looked to the particular properties of pattern generation and propagation and how these change our understanding of causality and agency. It argued that form is its own kind of real, one that emerges in the world and is amplified thanks to the distinctive manner in which humans and nonhumans harness it.

Chapter 6, "The Living Future (and the Imponderable Weight of the Dead)," turned to the afterlife of the spirit realm that lies beyond the realm of the living. Its primary task was to understand how this realm says something about the way life itself continues beyond the living bodies that breathe that life. (Spirit, I should note, is etymologically related to breath, and in Quichua, *samai*, breath, is what animates.) The last chapter, then, ventured beyond the existent into the "general." Generals are real; spirits, and even Sphinxes, are real. So are Lions. This chapter, then, is, one might say, about the reality of Lion as both kind and type. Lion as "kind" (or species, or lineage) is the product of life broadly construed, whereas Lion as "type" is the product of a human symbolic form of life. And this chapter focused on the emergent real that comes into being thanks to the particular ways these two kinds of generals— the living one beyond the human and the one that is distinctively human— come to be held together in the forest's ecology of selves.

This emergent real that comes into being in the forests around Ávila is the spirit realm of the masters. It is the product of a special configuration of concept and kind. It is a real that lies beyond the forest in ways that also catch up the life of the forest at the same time that it entangles that life with the all-too-human histories of the many dead that continue to haunt this forest that houses the masters.

Throughout this book I have sought ways to account for difference and novelty despite continuity. *Emergence* is a technical term I used to trace linkages across disjuncture; *beyond* is a broader, more general, one. That beyond human language lies semiosis reminds us that language is connected to the semiosis of the living world, which extends beyond it. That there are selves beyond the human draws attention to the fact that some of the attributes of our human selfhood are continuous with theirs. That there is a death beyond every life gestures toward the ways we might continue, thanks to the spaces opened up by all the absent dead who make us what we are. That form extends beyond life draws our attention to the effortless propagation of pattern that runs through our lives. And, finally, that spirits are a real part of an afterlife that extends beyond life tells us something about the continuity and generality intrinsic to life itself.

I hope to have provided here, in traversing this *selva selvaggia*, this wild "dense" and "difficult" forest where words so often fail us, some intimation of how it is that forests think. This thinking is amplified in a dense ecology of selves and certain historically contingent Runa ways of attending to that ecology.

Runa ways of attending to this forest ecology of selves are (in part) the product of an all-too-human marginalization from the national economy that might otherwise more equitably link rural communities like Ávila to some of Ecuador's growing wealth. Greater integration into national networks will certainly offer much more secure forms of sustenance, ones that would make the more onerous and riskier search for food in the forest obsolete and largely irrelevant. And things are moving in this direction. Quito is—through the nationwide expansion of roads, advances in health care, education, infrastructure, and so on—finally, after all these centuries, coming to the forest.

In pointing out the relation between socioeconomic and political marginalization and the forest-oriented subsistence that the Ávila Runa practice, I do not wish to reduce culture to poverty (as some would). Furthermore, I'm not, as should by now be clear, talking about culture. What is more, there is a certain plentitude to daily life in Ávila, one that is cherished by those who live in Ávila. And this richness exists regardless of the economic or health metrics one might use to evaluate it.

The particular colonially inflected, multispecies ecology of selves that I have described here is real in an ethnographic and ontological sense. But it depends for its existence on the continuous flourishing of dense nonhuman ecologies just as it does on the humans who live by tapping into those ecologies. If too many of these elements that make up this ecology of selves disappear, a particular kind of life (and afterlife) will come to an end—forever. And we will have to find ways to mourn its absence.

But it is not as if all life will end. There will be other Runa ways of being human—ones that might well also entangle nonhumans, ones that might call forth other spirits. And we must find ways to listen for the hopes that that kind of reality houses as well.

In turning my ethnographic attention to something potentially ephemeral and fleeting—the reality of a particularly dense ecology of selves, one that is both all too human and lies well beyond the human—I am not doing salvage anthropology. For what I am charting does not just disappear; ethnographic attention to this particular set of relations amplifies and thus allows us to appreciate ways of attending to the living logics that are already part of how forests think themselves through "us." And if "we" are to survive the Anthropocene—this indeterminate epoch of ours in which the world beyond the human is being increasingly made over by the all-too-human—we will have to actively cultivate these ways of thinking with and like forests.

I want, in this regard, to return to my title, *How Forests Think*. I chose this title because of its resonance, as I've noted, with Lévy-Bruhl's *How Natives Think*, a classic treatment of animistic thinking. At the same time, I wish to draw an important distinction: forests think; and when "natives" (or others, for that matter) think about that, they are made over by the thoughts of a thinking forest. My title *How Forests Think* also resonates with *La Pensée sauvage*, Lévi-Strauss's meditation on wild thought. Lévi-Strauss's meditation is about a kind of thought that both is and is not domesticated by the human. In this way it is like the ornamental flower the pansy—that other meaning of *pensée*—to which his title playfully alludes. Despite the fact that the pansy is domesticated, and therefore "tame," it is also alive. And thus, like us, and like the Runa—those *"indios mansos"*—the pansy is also wild. *Sauvage*, of course, is etymologically related to *sylvan*—that which is of the (wild) forest, the "selva selvaggia."

My own ethnographic meditation has been an attempt to liberate our thinking. It has been an attempt to step out, for a moment, of our doubt-ridden human housing to open ourselves to those wild living thoughts beyond the human—those that also make "us." To do this, we need to leave our guide the runa puma—our Virgil—and we also need to leave that forest, the selva selvaggia around Ávila. We do so, not necessarily to ascend to Dante's heavenly spheres (this is not that kind of morality tale; I'm not talking about that kind of telos). We leave this forest to step, for a moment, on our own, into a generality: one that is ethereal, perhaps, and one that lies beyond this particular ethnographic encounter.

In finding ways to open our thinking to living thoughts, to selves and souls, to the forest's many spirits, and even to the Lion as concept and kind, I have been trying to say something concrete about something general. I have been trying to say something about a general that makes itself felt in us "here" at the same time that it extends beyond us, over "there." Opening our thinking in this way might allow us to realize a greater *Us*—an *Us* that can flourish not just in our lives, but in the lives of those who will live beyond us. That would be our gift, however modest, to the living future.

NOTES

1. For my treatment of Quichua I adopt a practical orthography based on Spanish from Orr and Wrisley (1981: 154). In addition I use an apostrophe (" ' ") to indicate stops and a superscript h (" ^{h} ") to indicate aspiration. Words are to be stressed on the penultimate syllable unless indicated by an accent. The plural marker in Quichua is *-guna*. However, for reasons of clarity, I usually do not include the plural marker in my discussion of individual Quichua words even in contexts in which I use the term in its plural form in English. A hyphen ("-") indicates that word parts are suppressed. I use an en-dash ("–") to indicate where the vowels of a word have been drawn out. I use an em-dash ("—") to indicate an even greater elongation.

2. For ethnographic monographs on the Quichua-speaking Runa of Ecuador's Upper Amazon, see Whitten (1976), Macdonald (1979), and Uzendoski (2005). Muratorio (1987) and Oberem (1980) situate Runa lifeways within colonial and republican history and a broader political economy. For Ávila, see Kohn (2002b).

3. *Aya huasca* is prepared from a vine of the same name (*Banisteriopsis caapi*, Malpighiaceae) and sometimes mixed with other ingredients.

4. Norman Whitten's classic monograph, *Sacha Runa* (1976), astutely captures this tension between sylvan and civilized inherent to Runa ways of being.

5. All translations from Spanish and Quichua are my own.

6. In an earlier work (Kohn 2007) I referred to my approach as an "anthropology of life." The current iteration is closely related to that approach except that here I am less interested in the anthropological treatment of a subject matter (an anthropology of *x*) and more in an analytic that can take us beyond our subject matter ("the human") without abandoning it. Although so much of what we can learn about the human involves thinking with the logics of life that extend beyond the human, taking anthropology beyond the human also requires, as I will show, looking beyond life.

7. I do not deny the fact that certain "multinatural" forms of being in and understanding the world, including, most conspicuously, Amazonian ones, can shed critical light on what, by contrast, we can come to see as our folk academic "multicultural" conventions (Viveiros de Castro 1998). Nevertheless, the multiplication of natures is not an antidote to the problem posed by the multiplication of cultures.

8. A caffeine-rich beverage made from *Ilex guayusa* (Aquifoliaceae), a plant that is closely related to that used to make Argentinian mate.

9. I collected over 1,100 specimens of plants as well as 24 specimens of fungi. These are housed in the Herbario Nacional, Quito, with duplicates in the Missouri Botanical Garden. I also collected over 400 specimens of invertebrates, over 90 specimens of herpetofauna, and almost 60 specimens of mammals (all housed in the zoological museum of the Universidad Católica, Quito). My 31 specimens of fish are housed in the zoological museum of the Escuela Politécnica Nacional, Quito. Making specimens of birds is very difficult, requiring the complex preparation of skins. Therefore, I decided instead to document local avian faunal knowledge by taking close-up photographs of hunted specimens and conducting interviews using illustrated field manuals and recordings of calls.

10. By "relata," I mean a term, object, or entity that is constituted by its relationships to other such terms, objects, or entities, in the relational system in which it exists.

11. This form of citation, referring to the volume and paragraphs in Peirce's *Collected Papers* (1931), is the standard one used by Peirce scholars.

CHAPTER 1

1. I largely follow here the anthropological linguist Janis Nuckolls (1996) in her linguistic conventions for parsing Quichua. "Live" is an English gloss of the lexeme *causa-*; "2" indicates that it is conjugated for the second-person singular; "INTER" indicates that *-chu* is an interrogative, or question-marking suffix (see Cole 1985: 14–16).

2. In structuring my argument by asking you, the reader, to feel *tsupu*, I ask you to bracket, for a moment, your skepticism. But the argument still holds even if you don't "feel *tsupu*." As I will be discussing, *tsupu* exhibits formal properties (shared with similar sound images in all languages) that support the argument at hand (see also Sapir 1951 [1929]; Nuckolls 1999; Kilian-Hatz 2001).

3. I adopt "becoming worldly" from Donna Haraway (see Haraway 2008: 3, 35, 41) to invoke the possibility of inhabiting unprecedented and more hopeful emergent worlds through a practice of attention to those beings—human and nonhuman—that, in so many different ways, stand beyond us. Human language is both an impediment to and a vehicle for the realization of this project. This chapter attempts to explore how this is so.

4. From Marshall Sahlins's (1976: 12) classic anthropological statement on the relationship between culture and symbolic meaning to biology: "In the symbolic event, a radical discontinuity is introduced between culture and nature." This echoes Saussure's

(1959: 113) insistence on the "radically arbitrary" bond between "sound" (cf. nature) and "idea" (cf. culture).

5. This canopy emergent tree bearing big peapod-like fruits is known as *puca pacai* in Ávila (Latin *Inga alba*, Fabaceae-Mimosoideae).

6. See Kohn (2002b: 148–49) for the Quichua text.

7. For the purposes of this book I am collapsing a more complex division of the semiotic process, which, according to Peircean semiotics, involves three aspects: (1) a sign can be understood in terms of the characteristics it possesses in and of itself (whether it is a quality, an actual existent, or a law); (2) it can be understood in terms of the kind of relation it has to the object it represents; and (3) it can be understood in terms of the way its "interpretant" (a subsequent sign) represents it and its relation to its object. By using the term *sign vehicle* I am focusing here on the first of these three divisions. In general, however, as I will explain in the text, I am only treating signs as icons, indices, or symbols. In the process I am consciously collapsing the triadic division outlined above. Whether a sign is an icon, index, or symbol refers technically only to the second of the three divisions of the sign process (see Peirce CP 2.243–52).

8. Cf. Peirce's discussion of how suppression of certain features draws the attention to other ones in what he terms "diagrammatic icons" (Peirce 1998b: 13).

9. Of course the icon *pu oh* can also serve as an index (to be defined later in the text) at another level of interpretation. Like the event it is like, it can also startle someone who hears it.

10. See Peirce (1998d: 8).

11. See Peirce (CP 1.346, 1.339).

12. See Peirce (CP 1.339).

13. In this regard, note how in Peirce's pragmatism, "means" and "meaning" are related (CP 1.343).

14. See Peirce (CP 1.213).

15. Note that by recognizing how all signs, linguistic and otherwise, always "do things" we no longer need to appeal to a performative theory to make up for the deficiencies of a view of language as reference bereft of action (see Austin 1962).

16. See my discussion in the introduction on how even those anthropological approaches that recognize signs other than symbols still see these as exclusively human and interpretively framed by symbolic contexts.

17. Latin *Solanum quitoense*.

18. See Kohn (1992).

19. This example is adapted from Deacon's (1997: 75–76) discussion of iconism and the evolution of cryptic moth coloration.

20. The argument I make here about the logical relation of indexicality to iconicity follows and is adapted from Deacon (1997: 77–78).

21. Deacon is describing and semiotically reinterpreting the research of Sue Savage-Rumbaugh (see Savage-Rumbaugh 1986).

22. See also Peirce (CP 2.302) and Peirce (1998d: 10).

23. By "inferential," I mean that lineages of organisms constitute "guesses" about the environment. Via an evolutionary selective dynamic organisms come increasingly to "fit" their environment (see chapter 2).

24. This tends to be collapsed in anthropological treatments of Peirce. That is, thirdness tends to be seen only as a human symbolic attribute (see, e.g., Keane 2003: 414, 415, 420) rather than a property inherent to all semiosis and, in fact, to all regularity in the world.

25. "[The categories of firstness, secondness, and thirdness] suggest a way of thinking; and the possibility of science depends upon the fact that human thought necessarily partakes of whatever character is diffused through the whole universe, and that its natural modes have some tendency to be the modes of action of the universe" (Peirce CP 1.351).

26. And yet we must also recognize Descartes's insights about the "firstness" of feeling and of self. "I think therefore I am" loses its sense (and feeling) when it is applied to the plural or to the second or third person—just as only you—as an I—can feel *tsupu*.

27. See Kohn (2002b: 150–51) for Quichua text.

28. See Kohn (2002b: 45–46) for Quichua text.

29. Quichua *pishcu anga*.

30. See Kohn (2002b: 76) for Quichua text.

31. As such, it is related to *ticu*, which is used in Ávila to describe clumsy ambulation (see Kohn 2002b: 76).

32. See Bergson (1911: 97). Such a mechanistic logic is only possible because there is already a (whole) self outside the machine that designs or builds it.

33. "Huañuchi shami machacui."

34. Quichua *huaira machacui*; Latin *Chironius* sp.

35. See Whitten (1985) on this practice of severing the head from the snake's body and its potential symbolism.

36. Steve Feld's *Sound and Sentiment* (1990) is an instantiation of this; it is a book-long meditation on the symbolic structures through which the Kaluli (and, eventually, the anthropologist writing about them) come to feel an image.

CHAPTER 2

1. Spanish *barbasco*; Latin *Lonchocarpus nicou*; known in Ávila simply as *ambi*, poison.

2. See Kohn (2002b: 114–15) for Quichua text.

3. I adopt this phrase from Peirce (CP 1.221) and apply it to a broader range of phenomena.

4. See Roy Rappaport (1999: 1) for the position that the human species lives "in terms of meanings it must construct in a world devoid of intrinsic meaning but subject to physical law."

5. That I insist on the centrality of telos as an emergent property inherent to the "enchanted" living world that extends beyond the human puts me at odds with Jane Bennett's (2001) recent reappropriation of enchantment.

6. See Bateson (2000c, 2002); Deacon (1997); Hoffmeyer (2008); Kull et al. (2009).

7. Following Peirce's observations regarding "interpretants" in relation to the thoughts they represent, the organism-as-sign would be "identical ... though more developed" (CP 5.316) with respect to its progenitor's representation of the world.

8. For a list of some of the organisms that signal to the Runa the coming of the season when the leafcutter ants will fly or, in some cases, more specifically, the exact day when the reproductive ants will emerge, see Kohn (2002b: 99–101).

9. For a discussion of the specimens I collected of organisms found in association with leafcutter ants at the time when the winged reproductive ants emerged, see Kohn (2002b: 97–98).

10. On the kin terminology the Runa use to describe insects, see Kohn (2002b: 267).

11. *Carludovica palmata*, Cyclanthaceae (see Kohn 2002b: 457 n. 16).

12. People in Ávila continue to try to communicate with the ants and their colonies after they have been trapped (see Kohn 2002b: 103 for a discussion).

13. There is actually another layer of interaction among semiotic selves that causes amplification of the differences among soil conditions, which I've left out of the main text for the sake of clarity. Herbivores are themselves preyed upon by a second level of predators. If it weren't for this constraint, herbivore populations would grow unchecked, and the result would be unlimited herbivory on plants living in rich soils. With unlimited herbivory, the differences afforded by different soils would become irrelevant.

14. See Descola (1994) for an eloquent anti-reductionist critique of environmental determinism related to Amazonian soils and the ecological assemblages they sustain.

15. Here is how John Law and Annemarie Mol characterize nonhuman agency in ways that link it specifically to the relationality of human language:

> Within material semiotics, an entity counts as an actor if it makes a perceptible difference. Active entities are relationally linked with one another in webs. They make a difference to each other: they make each other be. Linguistic semiotics teaches that words give each other meaning. Material semiotics extends this insight beyond the linguistic and claims that entities give each other being: that they enact each other. (Law and Mol 2008: 58)

16. Later in this same passage (CP 1.314), Peirce links this ability to imagine ourselves into the being of another human with our ability to do the same with animals.

17. Quichua *manduru*; Latin *Bixa orellana*, Bixaceae; English annatto (see Kohn 2002b: 272–73 for a discussion of its use in Ávila).

18. *Procyon cancrivorus.*

19. This leads Viveiros de Castro (1998: 478) to conclude that there are many natures, each associated with the body-specific interpretive world of a particular kind of being; there is only one culture—in this case, that of the Runa. Accordingly, he refers to this way of thinking as "multinaturalism" and uses it as a critique of the multicultural logic (i.e., many cultures, one nature) typical of contemporary Western folk academic thought, especially in the guise of cultural relativism (cf. Latour 1993: 106; 2004: 48).

20. See Kohn (2002b: 108–41) for a more extensive discussion and many more examples of perspectivism in everyday Ávila life.

21. *Dactylomys dactylinus.*

22. For descriptions of these tree causeways, see Descola (1996: 157).

23. "*Saqui su.*"

24. For descriptions of this call, see Emmons (1990: 225).

25. This woman was already a grandmother, so this form of flirtatious joking was not seen to be threatening. Such jokes would not be made in reference to younger, recently married women.

26. *Renealmia* sp., Zingiberaceae.

27. Quichua *carachama*; Latin *Chaetostoma dermorynchon*, Loricariidae.

CHAPTER 3

1. "Isma tucus canga, puma ismasa isman."

2. A contraction of *ima shuti.*

3. "Cara caralla ichurin."

4. Quichua *yuyaihuan,* with the ability to think, judge, or react to circumstances.

5. Quichua *riparana,* to reflect on, attend to, or consider.

6. See Peirce (CP 2.654).

7. See Kohn (2002b: 349–54) for the Quichua transcription of Ventura's exchange with his father's puma.

8. See Kohn (2002b: 358–61) for Quichua text.

9. He uses the word *chita* (*chai* 'that' + *-ta* direct object marker)—i.e., *balarcani chita*—to refer to the wounded animal, instead of *pai* (the third-person pronoun used for an animate being regardless of gender or status as human).

10. On laughter as a way of fostering the sort of intimate sociability that Overing and Passes (2000) call "conviviality," see Overing (2000).

11. "Shican tucun."

12. "Runata mana llaquin." The verb *llaquina* means both sadness and love in Ávila. There is no specific word for love in Ávila Quichua, although there is in Andean Ecuadorian Quichua (*juyana*). In the Andean dialects with which I am familiar, *llaquina* means only sadness.

13. Also known as *aya buda* or *aya tulana.*

14. "Cai mishqui yacuta upingu."

15. "Shinaca yayarucu tiarangui, astalla shamunchi."

16. The place where the afterbirth is buried is known as the *pupu huasi,* the house of the afterbirth.

17. *Urera baccifera,* Urticaceae. This is closely related to the stinging nettles, which among other things are used to keep living beings away (by blocking the paths of dogs and toddlers). It is befitting of the phantasmal nature of the aya that a nonstinging variety of nettles is employed to ward it off (see Kohn 2002b: 275).

18. "Huaglin, singa taparin."

19. See Kohn (2002b: 214–15) for the Quichua text of Narcisa's narrative.

20. Cavell also asks whether the term might extend to our relations to nonhuman animals.

21. Quichua *"casariana alma."*

22. Quichua *"curuna."*

23. "Catina curunashtumandami ta' canisca."

24. See Bateson (2000b: 486–87); Haraway (2003: 50).

25. See Fausto (2007) for an extensive discussion of the ethnological implications of this dilemma in Amazonia.

26. What Fausto (2007) calls the "direction of predation" can change.

27. "Mana tacana masharucu puñun."

28. Also known as *gainari*; Paedarinae, Staphylinidae.

29. "Yumai pasapi chimbarin alma." See also Uzendoski (2005: 133).

30. See Kohn (2002b: 469 n. 95) for a list of these.

31. Also known as *buhya panga*, possibly *Anthurium* sect. *Pteromischum* sp. nov. (see Kohn 1992).

32. It is possible that this is due to unusually high vascular pressure.

33. See Kohn (2002b: 130–31) for Quichua text.

34. See Kohn (2002b: 132) for Quichua text.

35. *Cedrelinga cateniformis*, Fabaceae-Mimosoideae.

36. See Kohn (2002b: 136–39) for Quichua text of this myth.

CHAPTER 4

1. This is a variant of *aya—i* discussed in chapter 2.

2. The term *all too human* alludes vaguely to Nietzsche (Nietzsche and Hollingdale 1986) and Weber (1948b: 132, 348). I develop the specific way I use it in the passages that follow.

3. Value has been the subject of lively discussion in anthropology. In large part this has centered on how to reconcile the various forms that value takes in human realms (see esp. Graeber 2001; see also Pederson 2008 and Kockelman 2011 for attempts to reconcile anthropological and economic theories of value with Peircean ones). My contribution to this literature is to stress the point that human forms of value stand in a relation of emergent continuity with a basic form of value that emerges with life.

4. See, in this regard, Coppinger and Coppinger (2002) on canine self-domestication.

5. See also Ellen (1999: 66); Haraway (2003: 41).

6. The main ingredient is the inner bark scrapings of the understory tree tsita (*Tabernaemontana sananho*, Apocynaceae). Other ingredients include tobacco and *lumu cuchi huandu* (*Brugmansia* sp., Solanaceae), a special canine variety of a very powerful belladonna-related narcotic sometimes used by Runa shamans.

7. Dogs partake of the following human qualities:

1. Unlike animals they are expected to eat cooked food.

2. According to some, they have souls that are capable of ascending to the Christian heaven.

3. They acquire the dispositions of their masters; mean owners have mean dogs.

4. Dogs and children who become lost in the forest become "wild" (Quichua *quita*) and therefore frightened of people.

8. See Oberem (1980: 66); see also Schwartz (1997: 162–63); Ariel de Vidas (2002: 538).

9. In fact mythic man-eating jaguars are said to refer to humans as palm hearts.

10. See Fausto (2007); Conklin (2001).

11. These are known in Ávila as "forest masters" (*sacha amuguna*) or "forest lords" (*sacha curagaguna*).

12. Colonial categories used historically to describe the Runa, such as Christian and *manso* (tame; Quichua *mansu*), as opposed to infidel (*auca*) and wild (*quita*), however problematic (see Uzendoski 2005: 165), cannot be discounted because, in Ávila at least, they currently constitute the idiom through which a certain kind of agency, albeit one that is not so overtly visible, is manifested (see chapter 6).

13. I thank Manuela Carneiro da Cunha for reminding me of this fact, to which several Ávila oral histories that I have collected attest. See also Blomberg (1957) for eyewitness written accounts and photographs of such expeditions.

14. The term *runa* is also used in Ecuadorian Spanish to describe cattle that are not an identifiable breed. It is also used to describe anything that is considered pejoratively as having supposedly "Indian" qualities (e.g., items considered shabby or dirty).

15. See also Haraway (2003: 41, 45).

16. Descola, regarding the Achuar, refers to this form of isolation as the "solipsism of natural idioms" (1989: 443). The emphasis he gives to the failure in communication thus implied is appropriate given this chapter's subject matter.

17. Willerslev's (2007) discussion of Siberian Yukaghir hunting treats in great detail this threat to human identity posed by relations with animals. The solutions the Yukaghir find are different; the general problem—the challenge of living socially in a world peopled by many kinds of selves—is the same.

18. Quichua *duiñu*, from the Spanish *dueño*.

19. For examples of this canine lexicon, see Kohn (2007: 21 n. 30).

20. As in chapter 1 I follow in this chapter Nuckolls (1996) in her linguistic conventions for parsing Quichua. These include the following: ACC = Accusative case; COR = Coreference; FUT = Future; NEG IMP = Negative Imperative; SUB = Subjunctive; 2 = Second person; 3 = Third person.

21. *Ucucha* refers to the class of small rodents that includes mice, rats, spiny rats, and mouse opossums. It is a euphemism for *sicu*, the class of large edible rodents that includes the agouti, paca, and agouchy.

22. Here is another example from Ávila, not discussed in the body of this chapter, of giving advice to dogs using canine imperatives while administering tsita:

2.1 *tiutiu-nga ni-sa*
chase-3FUT say-COR
thinking/desiring it will chase

2.2 *ama runa-ta capari-nga ni-sa*
NEG IMP person-ACC bark-3FUT say-COR
thinking/desiring it will not bark at people

23. I thank Bill Hanks for suggesting this term.

24. Regarding the anomalous use of a negative imperative in combination with a third-person future marker in line 1.2 (cf. lines 1.5 and 5.3 in the text and 2.2 in note 22), the following are related constructions that *would* be considered grammatically correct in everyday Ávila Quichua:

If addressed to a dog in the second person:

3 *atalpa-ta ama cani-y-chu*
chicken-ACC NEG IMP bite-2-IMP-NEG
don't bite chickens

If addressed to another person about a dog:

4a *atalpa-ta mana cani-nga-chu*
chicken-ACC NEG bite-3FUT-NEG
it will not bite chickens

or

4b *atalpa-ta ama cani-chun*
chicken-ACC NEG bite-SUB
so that it doesn't bite chickens

25. Regarding how humans can bring out human subjectivities in animals by denying them their bodies, compare reports and legends of Runa men undressing themselves before fighting off jaguars they encounter in the forest. By doing so, they remind jaguars that beneath their feline bodily habitus, which can be "divested" like clothing, they too are humans (see chapter 6).

26. According to Janis Nuckolls, Quichua speakers from the Pastaza region of Amazonian Ecuador refer to or address these spirits in songs using third-person future constructions (pers. com.). This is another reason for suspecting that the use of "señora" to address spirit lovers in Ávila is related to the use of "canine imperatives."

27. Reduplication is frequently used in imitating birdcalls and in onomatopoeic bird names in Ávila (see also Berlin and O'Neill 1981; Berlin 1992).

28. See also Taylor (1996); Viveiros de Castro (1998).

29. On distributed selfhood, see Peirce (CP 3.613; 5.421; 7.572). See also Strathern (1988: 162); and for a somewhat different take Gell (1998).

30. For the semiotic constraints of extraterrestrial grammars, see Deacon (2003).

CHAPTER 5

1. On how the Huaorani treat peccaries as social others, see Rival (1993).

2. Other examples of apparently spontaneous recognition of wild/domestic parallels by outsiders include the following:

1) Simson's (1878: 509) musings, elsewhere, about how his Záparo guides in Iquitos might compare the European horse with the tapir. In Ávila, the tapir, distant relative of the horse and the New World's only extant native odd-toed ungulate, is understood to be the horse of certain spirit masters of the forest.

2) The correspondence between white domestication and Indian forest predation as noted by the seventeenth-century Jesuit priest Figueroa who marveled at the nuts and fruits that "nature, like an orchard, provides" Amazonians and referred to the "herds of wild pigs" and other animals of the forest as Amazonian "livestock" ("crías") "that need no care" (Figueroa 1986 [1661]: 263).

3) The nineteenth-century Jesuit priest Pozzi who in a sermon in Loreto compared Runa hunting to civilized animal husbandry (in Jouanen 1977: 90).

3. See Janzen (1970); Wills et al. (1997).

4. My argument about the ways in which the rubber economy was formally constrained is at odds with, but ultimately not inconsistent with, what Steven Bunker has written. Bunker (1985: 68–69) argued that the fungal parasite is not enough to make rubber cropping in the Amazon impossible. Successful grafting and close planting techniques were developed in the Amazon, but these are labor-intensive, and what was lacking in this region was labor. Labor shortages, not parasites, according to Bunker, were what prevented plantation cropping. Surely, the form-propagating tendencies that the rubber boom reveals are weak ones, and with sufficient labor they might well become dampened or even irrelevant. But the shortage of labor at this time allowed for certain formal properties to become amplified and to propagate across a variety of domains, and to thus play a central role in the rubber economy.

5. *Salminus hilarii.*

6. *Virola duckei*, Myristicaceae.

7. For a description of rubber tapping and initial processing and the skill and effort required to get latex to rivers, see Cordova (1995).

8. See Irvine (1987) on the San José Runa preference for erecting hunting blinds by fruiting trees as opposed to searching for game in the forest. This is also a popular technique in Ávila. By waiting by a fruiting tree, hunters in effect harness floristic form.

9. See Oberem (1980: 117); Muratorio (1987: 107). For information on communities that descend from Ávila Runa forcibly resettled on the Peruvian Napo during the rubber boom, see Mercier (1979).

10. For another example of shamanistic harnessing of the Amazon riverine network, see Descola (1996: 323). See Kohn (2002a: 571–73) for an example of the ways in which Jesuit missionaries imagined the Amazon riverine network as a conduit for consecration and conversion.

11. See Martín (1989 [1563]: 119); Ordóñez de Cevallos (1989 [1614]: 429); Oberem (1980: 225).

12. See Oberem (1980: 117); Muratorio (1987); Gianotti (1997).

13. In contrast to other extractive products, such as minerals or petroleum, there is something unique to how certain life forms like wild Amazonian rubber (or wild mat-sutake mushrooms; see Tsing [2012]) can become commodities. Extraction of these, even under the most ruthless capitalist systems, requires entering into and, to an extent, succumbing to the relational logic that supports this living wealth. The aspect of that logic that concerns me here involves its patterned quality.

14. On the logical properties of hierarchy, see Bateson (2000e).

15. This sort of relationship of bird name to call is common in Ávila (see Kohn 2002b: 146 for another example).

16. *mashuta micusa sacsa rinu-*

17. *-napi imata cara*

18. In Descola's (2005) terms, Silverman's project is to trace the hidden modes of "analogic" thinking in a Western thought otherwise dominated by "naturalistic" thinking.

19. By "history" here, I mean our experience of the effect of past events on the present. Peirce refers to this as our experience of secondness, which includes our experience of change, difference, resistance, otherness, and time (CP 1.336; 1.419); see chapter 1. This is not to deny that there are specific and highly variable sociohistorically situated modalities of representing the past (see Turner 1988) or ideas about causality (Keane 2003). I am making a broader and more general set of claims, namely: (1) the experience of secondness is not necessarily delimited culturally; and (2) there are moments when the dyadic effects of the past on the present that we associate with history becomes less relevant as a causal modality.

20. By "time," I mean the directional process spanning from past to present to probable future. I am making no absolute claim about the ontological status of time. Neither, however, do I want to say that time is wholly a cultural or even a human construct (cf. CP 8.318). My argument is at the level of what Bateson calls "creatura" (2000a: 462). That is, in the realm of life, the past, present, and probable each comes to have specific properties, and these properties are intimately involved in the ways in which semiotic selves represent the world around them. For it is in the realm of life, via semiosis, that the future comes to affect the present through the vehicle of representation (see Peirce CP 1.325). See also chapter 6.

21. Both glossed in Quichua as *turmintu* (from the Spanish *tormento*).

22. In the spirit master's realm, they escape Judgment Day, *juiciu punja*.

23. See Peirce (CP 6.101).

24. Jonathan Hill (1988) and several other contributors to his edited volume provide a critique of Lévi-Strauss's hot/cold distinction. Hill argues that this distinction erases the many ways in which Amazonians are products of, producers of, and conscious of history. Peter Gow (2001) has argued that such a critique misses Lévi-Strauss's point: myths are responses to history in that they are, as Gow puts it,

"instruments for the obliteration of time" (27). That myths have this characteristic is evident. What is less clear from Gow's analysis is why. My argument is that timelessness is an effect of the peculiar properties of form.

25. Cf. Lévi-Strauss: "odds and ends left over from psychological or historical processes, . . . [which] appear as such only in relation to the history which produced them and not from the point of view of the logic for which they are used" (1966: 35).

26. For the position that Amazonian landscape and natural history are always in some ways social, see Raffles (2002). On the "pristine myth" and for a review of the literature on anthropogenic forests, see Denevan (1992); Cleary (2001). Without denying the importance of historicizing "natural history," the position I take is somewhat different. The idea that all nature is always already historical is related to the representational problem we face in our field—namely, that we don't know how to talk about that which stands outside the human-specific conventional logic of symbolic reference without reducing the human to matter (see chapter 1).

27. On the hopes for symmetrical relations between Upper Amazonians and Europeans, see Taylor (1999: 218).

28. See Kohn (2002b: 363–64) for a more detailed account.

29. People in Ávila today recount a myth that explains why a certain king, sometimes referred to as an Inca, abandoned his attempts to build Quito near Ávila and finally built it in the Andes. Some people even discern the remnants of this failed jungle Quito in the landscape. This idea of a Quito having quite literally abandoned the region also comes up in the nearby community Oyacachi (see Kohn 2000b: 249–50; Kohn 2002a).

30. There are also all-too-human contexts in which form propagates. Late Soviet socialism provides one such example (see Yurchak 2006, 2008; and my comment on the latter [Kohn 2008]). Here, the severing of official discursive form from any indexical specification—a form that was nevertheless sustained by the entire might of the Soviet state—allowed a certain kind of invisible self-organizing politics to emerge spontaneously and simultaneously throughout various parts of the Soviet Union. Yurchak appropriately calls this a "politics of indistinction," alluding to the way it harnessed and proliferated official discursive forms (for some sort of an end, however undefined) rather than acquiescing to or resisting them.

31. See Peirce (1998d: 4); cf. Bateson (2000d: 135).

32. Quoted in Colapietro (1989: 38). I thank Frank Salomon for first drawing my attention to this passage.

CHAPTER 6

1. Quichua *sahinu chuspi* (peccary flies); Latin Diptera.

2. By drawing on Freud's understanding of the uncanny, as "that species of the frightening that goes back to what was once well known and had long been familiar" (Freud 2003: 124), I wish to make explicit reference to Mary Weismantel's (2001) treatment of the *pishtaco*, the white bogeyman of the Andes that eats Indian fat. The pish-

taco, like the policeman for Oswaldo, is inextricably embedded in what it is to be Andean in ways that are uncanny—frightening but also intimate and familiar.

3. And yet such a generalized power could not exist without the specific instances of its manifestation. Structures of domination are ultimately given their "brutal" efficacy through what Peirce has called "secondness" (see chapter 1), manifest, according to one example he gives, in "the sheriff's hand" on your shoulder (CP 1.24) or, in Oswaldo's case, in the policeman who suddenly appears at a friend's door (see CP 1.213). And yet, as Butler underscores, power is something more than such easily externalized brutality.

4. We live in a sort of gift economy with the dead, with the spirits, and with the future selves we might come to be and without whom we are nothing. Marcel Mauss's notion of the debts that make us who we are applies to our relations to all of these others: "by giving one is giving oneself, and if one gives oneself, it is because one 'owes' oneself—one's person and one's goods—to others" (Mauss 1990 [1950]: 46).

5. Wooden slit drums used for long-distance communication were among the first things that the Spaniards banned in the Upper Amazon (Oberem 1980).

6. This is not to say that they would consider themselves unclothed. Penis strings and face paints function in important ways as clothing.

7. "hacerlos de brutos, hombres, y de hombres, cristianos."

8. This form of always already inhabiting something that might otherwise be understood as the cumulative effect of history makes itself manifest in Oyacachi, a cloud forest village to the west of Ávila that in the early colonial period was part of the same Quijos chiefdom. As people there understand it there was never a time when they were not Christian. In fact, according to one myth (see Kohn 2002a), it is the white European priests, not the natives, who are the pagans in need of conversion.

9. Sometimes, of course, self-objectification is an important strategy for achieving political visibility.

10. Refillable metal shotgun cartridges have a little hole at the base where one fits the firing cap. Oswaldo's dream image, I should note, has shamanistic overtones. Blowing through a shotgun cartridge is like blowing through a blowgun, and sorcerers attack their victims by placing their cupped hands to their mouths and shooting invisible blowgun darts (sagra tullu) at their victims.

11. By "topic" here, I mean the theme of a sentence, that about which the sentence gives information, as opposed to its grammatical subject, which may or may not also be the topic. Quichua speakers often mark the topic (which may be either the subject, object, adverb, or verb of the sentence) for a number of reasons, including, as is done in the example treated here, to emphasize a theme that might not otherwise be noted given the assumed context. For a discussion of topic, on which my treatment of the matter is based, and for a further explanation of the use of topic-marking suffixes in Ecuadorian Quichua, see Chuquín and Salomon (1992: 70–73) and Cole (1985: 95–96).

12. For the Quichua text, see Kohn (2002b: 292).

13. In an otherwise identical series of myths these apostles replace the well-known culture hero brothers Cuillur and Duciru of other Upper Amazonian Runa communities (e.g., Orr and Hudelson 1971).

14. Urban writes about this in terms of the continuity of "culture," not of self.

15. "In the flow of time in the mind, the past appears to act directly upon the future, its effect being called memory, while the future only acts upon the past through the medium of thirds" (CP 1.325).

16. This is in reference to the Tupian Omagua.

17. See Gianotti (1997: 128); Oberem (1980: 290); Wavrin (1927: 335).

18. See Wavrin (1927: 335); see also Gianotti (1997: 128); Avendaño (1985 [1861]: 152); Orton (1876: 193); Colini (1883: 296); cf. Maroni (1988 [1738]: 172, 378); Kohn (2002b: 238).

19. "Chuchuyu," "with breasts," was how Ventura referred to Rosa's granddaughters, before explaining that in the master's realm, Rosa would "live forever, never to die again, without suffering, like a child" ("Huiñai huiñai causangapa, mana mas huañungapa, mana tormento, huahuacuintallata").

20. This probably refers to the barred antshrike.

21. Quichua: runa pamba (lit., "people burier"); English: tarantula hawk; Latin: Pepsis sp., Pompilidae.

22. For more such examples, see Kohn (2002b: 242–43, 462 n. 54).

23. Ginsberg's "kaddish" does mention death.

24. For a discussion of names for shamans and shamanism, see Kohn (2002b: 336–38).

25. Regarding the abandonment of shorts for long pants among the Tena Runa, see Gianotti (1997: 253).

26. Wavrin similarly reports that men who encounter jaguars are not afraid of them and can do battle with them, "fighting one-on-one as equals" as if they were men because they know these jaguars were once men (Wavrin 1927: 335; see also Kohn 2002b: 270).

27. Cushma refers to a gown traditionally worn by Cofán as well as western Tukanoan Siona and Secoya men.

28. See Kohn (2002b: 271–72) for an early colonial Ávila area example of the use of clothing to confer power.

29. "Pucuhuai, camba yachaita japingapa."

30. See Kohn (2002b: 281) for eighteenth-century Amazonian strategies of appropriating white clothing as equipment.

31. My thinking about survival has been greatly influenced by Lisa Stevenson's work.

BIBLIOGRAPHY

Agamben, Giorgio
  2004   The Open: Man and Animal. Stanford, CA: Stanford University Press.
Ariel de Vidas, Anath
  2002   A Dog's Life among the Teenek Indians (Mexico): Animals' Participation in the Classification of Self and Other. Journal of the Royal Anthropological Institute, n.s., 8: 531–50.
Austin, J. L.
  1962   How to Do Things with Words. Oxford: Clarendon Press.
Avendaño, Joaquín de
  1985 [1861]   Imagen del Ecuador: Economía y sociedad vistas por un viajero del siglo XIX. Quito, Ecuador: Corporación Editora Nacional.
Bateson, Gregory
  2000a   Form, Substance, and Difference. In Steps to an Ecology of Mind. Pp. 454–71. Chicago: University of Chicago Press.
  2000b   Pathologies of Epistemology. In Steps to an Ecology of Mind. G. Bateson, ed. Pp. 486–95. Chicago: University of Chicago Press.
  2000c   Steps to an Ecology of Mind. Chicago: University of Chicago Press.
  2000d   Style, Grace, and Information in Primitive Art. In Steps to an Ecology of Mind. G. Bateson, ed. Pp. 128–52. Chicago: University of Chicago Press.
  2000e   A Theory of Play and Fantasy. In Steps to an Ecology of Mind. G. Bateson, ed. Pp. 177–93. Chicago: University of Chicago Press.
  2002   Mind and Nature: A Necessary Unity. Creskill, NJ: Hampton Press.
Bennett, Jane
  2001   The Enchantment of Modern Life: Attachments, Crossings, and Ethics. Princeton, NJ: Princeton University Press.
  2010   Vibrant Matter: A Political Ecology of Things. Durham, NC: Duke University Press.

Benveniste, Émile
   1984   The Nature of Pronouns. *In* Problems in General Linguistics. E. Benveniste,
          ed. Pp. 217–22. Coral Gables, FL: University of Miami Press.
Berger, John
   2009   Why Look at Animals? London: Penguin.
Bergson, Henri
   1911   Creative Evolution. New York: H. Holt and Co.
Berlin, Brent
   1992   Ethnobiological Classification: Principles of Categorization of Plants
          and Animals in Traditional Societies. Princeton, NJ: Princeton University
          Press.
Berlin, Brent, and John P. O'Neill
   1981   The Pervasiveness of Onomatopoeia in Aguaruna and Huambisa Bird
          Names. Journal of Ethnobiology 1 (2): 238–61.
Blomberg, Rolf
   1957   The Naked Aucas: An Account of the Indians of Ecuador. F. H. Lyon, trans.
          Fair Lawn, NJ: Essential Books.
Borges, Luis
   1998   Funes, the Memorious. *In* Fictions. A. Kerrigan, ed. Pp. 97–105. London:
          Calder Publications.
Brockway, Lucile
   1979   Science and Colonial Expansion: The Role of the British Royal Botanic
          Gardens. New York: Academic Press.
Buber, Martin
   2000   I and Thou. New York: Scribner.
Bunker, Stephen G.
   1985   Underdeveloping the Amazon: Extraction, Unequal Exchange, and the Fail-
          ure of the Modern State. Urbana: University of Illinois Press.
Butler, Judith
   1997   The Psychic Life of Power: Theories in Subjection. Stanford, CA: Stanford
          University Press.
Camazine, Scott
   2001   Self-Organization in Biological Systems. Princeton, NJ: Princeton Univer-
          sity Press.
Campbell, Alan Tormaid
   1989   To Square with Genesis: Causal Statements and Shamanic Ideas in Wayãpí.
          Iowa City: University of Iowa Press.
Candea, Matei
   2010   Debate: Ontology Is Just Another Word for Culture. Critique of Anthropol-
          ogy 30 (2): 172–79.
Capps, Lisa, and Elinor Ochs
   1995   Constructing Panic: The Discourse of Agoraphobia. Cambridge, MA: Har-
          vard University Press.

Carrithers, Michael
 2010   Debate: Ontology Is Just Another Word for Culture. Critique of Anthropology 30 (2): 156–68.
Cavell, Stanley
 2005   Philosophy the Day after Tomorrow. Cambridge, MA: Belknap Press.
 2008   Philosophy and Animal Life. New York: Columbia University Press.
Chakrabarty, Dipesh
 1997   The Time of History and the Times of Gods. In The Politics of Culture in the Shadow of Capital. L. Lowe and D. Lloyd, eds. Pp. 35–60. Durham, NC: Duke University Press.
 2000   Provincializing Europe: Postcolonial Thought and Historical Difference. Princeton, NJ: Princeton University Press.
Choy, Timothy K., et al.
 2009   A New Form of Collaboration in Cultural Anthropology: Matsutake Worlds. American Ethnologist 36 (2): 380–403.
Chuquín, Carmen, and Frank Salomon
 1992   Runa Shimi: A Pedagogical Grammar of Imbabura Quichua. Madison: Latin American and Iberian Studies Program, University of Wisconsin–Madison.
Cleary, David
 2001   Toward an Environmental History of the Amazon: From Prehistory to the Nineteenth Century. Latin American Research Review 36 (2): 64–96.
Colapietro, Vincent M.
 1989   Peirce's Approach to the Self: A Semiotic Perspective on Human Subjectivity. Albany: State University of New York Press.
Cole, Peter
 1985   Imbabura Quechua. London: Croom Helm.
Colini, G. A.
 1883   Collezione Etnologica degli Indigeni dell' Alto Amazzoni Acquistata dal Museo Preistorico-Etnografico di Roma. Bollettino della Società Geografica Italiana, anno XVII, vol. XX, ser. II; vol. VIII: 287–310, 353–83.
Conklin, Beth A.
 2001   Consuming Grief: Compassionate Cannibalism in an Amazonian Society. Austin: University of Texas Press.
Conklin, Beth A., and Laura R. Graham
 1995   The Shifting Middle Ground: Amazonian Indians and Eco-Politics. American Anthropologist 97 (4): 695–710.
Coppinger, Raymond, and Lorna Coppinger
 2002   Dogs: A New Understanding of Canine Origin, Behavior, and Evolution. Chicago: University of Chicago Press.
Cordova, Manuel
 1995   Amazonian Indians and the Rubber Boom. In The Peru Reader: History, Culture, Politics. O. Starn, C. I. Degregori, and R. Kirk, eds. Pp. 203–14. Durham, NC: Duke University Press.

Csordas, Thomas J.
  1999   The Body's Career in Anthropology. *In* Anthropological Theory Today. H. L. Moore, ed. Pp. 172–205. Cambridge: Polity Press.
Cunha, Manuela Carneiro da
  1998   Pontos de vista sobre a floresta amazônica: Xamanismo e tradução. Mana 4 (1): 7–22.
Daniel, E. Valentine
  1996   Charred Lullabies: Chapters in an Anthropology of Violence. Princeton, NJ: Princeton University Press.
de la Cadena, Marisol
  2010   Indigenous Cosmopolitics in the Andes: Conceptual Reflections beyond "Politics." Cultural Anthropology 25 (2): 334–70.
de Ortiguera, Toribio
  1989 [1581–85]   Jornada del río Marañon, con todo lo acaecido en ella y otras cosas notables dignas de ser sabidas, acaecidas en las Indias occiden-tales. . . *In* La Gobernación de los Quijos (1559–1621). C. Landá-zuri, ed. Pp. 357–80. Iquitos, Peru: IIAP-CETA.
Deacon, Terrence W.
  1997   The Symbolic Species: The Co-evolution of Language and the Brain. New York: Norton.
  2003   The Hierarchic Logic of Emergence: Untangling the Interdependence of Evolution and Self-Organization. *In* Evolution and Learning: The Baldwin Effect Reconsidered. B. Weber and D. Depew, eds. Pp. 273–308. Cambridge, MA: MIT Press.
  2006   Emergence: The Hole at the Wheel's Hub. *In* The Re-Emergence of Emergence: The Emergentist Hypothesis from Science to Religion. P. Clayton and P. Davies, eds. Pp. 111–50. Oxford: Oxford University Press.
  2012   Incomplete Nature: How Mind Emerged from Matter. New York: Norton.
Dean, Warren
  1987   Brazil and the Struggle for Rubber: A Study in Environmental History. Cambridge: Cambridge University Press.
Deleuze, Gilles, and Félix Guattari
  1987   A Thousand Plateaus: Capitalism and Schizophrenia. Minneapolis: University of Minnesota Press.
Denevan, William M.
  1992   The Pristine Myth: The Landscape of the Americas in 1492. Annals of the Association of American Geographers 82 (3): 369–85.
Dennett, Daniel Clement
  1996   Kinds of Minds: Toward an Understanding of Consciousness. New York: Basic Books.
Derrida, Jacques
  2008   The Animal That Therefore I Am. New York: Fordham University Press.

Descola, Philippe
  1989   Head-Shrinkers versus Shrinks: Jivaroan Dream Analysis. Man, n.s., 24: 439–50.
  1994   In the Society of Nature: A Native Ecology in Amazonia. Cambridge: Cambridge University Press.
  1996   The Spears of Twilight: Life and Death in the Amazon Jungle. New York: New Press.
  2005   Par-delà nature et culture. Paris: Gallimard.
Diamond, Cora
  2008   The Difficulty of Reality and the Difficulty of Philosophy. *In* Philosophy and Animal Life. Stanley Cavell et al., eds. Pp. 43–89. New York: Columbia University Press.
Duranti, Alessandro, and Charles Goodwin
  1992   Rethinking Context: Language as an Interactive Phenomenon. Cambridge: Cambridge University Press.
Durkheim, Émile
  1972   Selected Writings. Cambridge: Cambridge University Press.
Ellen, Roy
  1999   Categories of Animality and Canine Abuse: Exploring Contradictions in Nuaulu Social Relationships with Dogs. Anthropos 94: 57–68.
Emerson, Ralph Waldo
  1847   The Sphinx. *In* Poems. Boston: James Munroe and Co.
Emmons, Louise H.
  1990   Neotropical Rainforest Mammals: A Field Guide. Chicago: University of Chicago Press.
Evans-Pritchard, E. E.
  1969   The Nuer: A Description of the Modes of Livelihood and Political Institutions of a Nilotic People. Oxford: Oxford University Press.
Fausto, Carlos
  2007   Feasting on People: Eating Animals and Humans in Amazonia. Current Anthropology 48 (4): 497–530.
Feld, Steven
  1990   Sound and Sentiment: Birds, Weeping, Poetics, and Song in Kaluli Expression. Philadelphia: University of Pennsylvania Press.
Figueroa, Francisco de
  1986 [1661]   Informes de Jesuitas en el Amazonas. Iquitos, Peru: IIAP-CETA.
Fine, Paul
  2004   Herbivory and Evolution of Habitat Specialization by Trees in Amazonian Forests. PhD dissertation, University of Utah.
Fine, Paul, Italo Mesones, and Phyllis D. Coley
  2004   Herbivores Promote Habitat Specialization in Trees in Amazonian Forests. Science 305: 663–65.

Foucault, Michel
    1970   The Order of Things: An Archaeology of the Human Sciences. London: Tavistock.
Freud, Sigmund
    1965   The Psychopathology of Everyday Life. J. Strachey, trans. New York: Norton.
    1999   The Interpretation of Dreams. Oxford: Oxford University Press.
    2003   The Uncanny. H. Haughton, trans. London: Penguin.
Gell, Alfred
    1998   Art and Agency: An Anthropological Theory. Oxford: Clarendon Press.
Gianotti, Emilio
    1997   Viajes por el Napo: Cartas de un misionero (1924–1930). M. Victoria de Vela, trans. Quito, Ecuador: Ediciones Abya-Yala.
Ginsberg, Allen
    1961   Kaddish, and Other Poems, 1958–1960. San Francisco: City Lights Books.
Gow, Peter
    1996   River People: Shamanism and History in Western Amazonia. In Shamanism, History, and the State. C. Humphrey and N. Thomas, eds. Pp. 90–113. Ann Arbor: University of Michigan Press.
    2001   An Amazonian Myth and Its History. Oxford: Oxford University Press.
Graeber, David
    2001   Toward an Anthropological Theory of Value: The False Coin of Our Own Dreams. New York: Palgrave.
Hage, Ghassan
    2012   Critical Anthropological Thought and the Radical Political Imaginary Today. Critique of Anthropology 32 (3): 285–308.
Haraway, Donna
    1999   Situated Knowledges: The Science Question in Feminism and the Privilege of Partial Perspective. In The Science Studies Reader. M. Biagioli, ed. Pp. 172–201. New York: Routledge.
    2003   The Companion Species Manifesto: Dogs, People, and Significant Otherness. Chicago: Prickly Paradigm Press.
    2008   When Species Meet. Minneapolis: University of Minnesota Press.
Hare, Brian, et al.
    2002   The Domestication of Social Cognition in Dogs. Science 298: 1634–36.
Hemming, John
    1987   Amazon Frontier: The Defeat of the Brazilian Indians. London: Macmillan.
Hertz, Robert
    2007   The Pre-eminence of the Right Hand: A Study in Religious Polarity. In Beyond the Body Proper. M. Lock and J. Farquhar, eds. Pp. 30–40. Durham, NC: Duke University Press.
Heymann, Eckhard W., and Hannah M. Buchanan-Smith
    2000   The Behavioural Ecology of Mixed Species of Callitrichine Primates. Biological Review 75: 169–90.

Hill, Jonathan D.
  1988   Introduction: Myth and History. *In* Rethinking History and Myth: Indigenous South American Perspectives on the Past. J.D. Hill, ed. Pp. 1–18. Urbana: University of Illinois Press.
Hilty, Steven L., and William L. Brown
  1986   A Guide to the Birds of Colombia. Princeton, NJ: Princeton University Press.
Hoffmeyer, Jesper
  1996   Signs of Meaning in the Universe. Bloomington: Indiana University Press.
  2008   Biosemiotics: An Examination into the Signs of Life and the Life of Signs. Scranton, PA: University of Scranton Press.
Hogue, Charles L.
  1993   Latin American Insects and Entomology. Berkeley: University of California Press.
Holbraad, Martin
  2010   Debate: Ontology Is Just Another Word for Culture. Critique of Anthropology 30 (2): 179–85.
Hudelson, John Edwin
  1987   La cultura quichua de transición: Su expansión y desarrollo en el Alto Amazonas. Quito: Museo Antropológico del Banco Central del Ecuador (Guayaquil), Ediciones Abya-Yala.
Ingold, Tim
  2000   The Perception of the Environment: Essays in Livelihood, Dwelling and Skill. London: Routledge.
Irvine, Dominique
  1987   Resource Management by the Runa Indians of the Ecuadorian Amazon. Ph.D. dissertation, Stanford University.
Janzen, Daniel H.
  1970   Herbivores and the Number of Tree Species in Tropical Forests. American Naturalist 104 (904): 501–28.
  1974   Tropical Blackwater Rivers, Animals, and Mast Fruiting by the Dipterocarpaceae. Biotropica 6 (2): 69–103.
Jiménez de la Espada, D. Marcos
  1928   Diario de la expedición al Pacífico. Boletín de la Real Sociedad Geográfica 68 (1–4): 72–103, 142–93.
Jouanen, José
  1977   Los Jesuítas y el Oriente ecuatoriano (Monografía Histórica), 1868–1898. Guayaquil, Ecuador: Editorial Arquidiocesana.
Keane, Webb
  2003   Semiotics and the Social Analysis of Material Things. Language and Communication 23: 409–25.

Kilian-Hatz, Christa
    2001   Universality and Diversity: Ideophones from Baka and Kxoe. *In* Ideophones. F. K. E. Voeltz and C. Kilian-Hatz, eds. Pp. 155–63. Amsterdam: John Benjamin.

Kirksey, S. Eben, and Stefan Helmreich
    2010   The Emergence of Multispecies Ethnography. Cultural Anthropology 25 (4): 545–75.

Kockelman, Paul
    2011   Biosemiosis, Technocognition, and Sociogenesis: Selection and Significance in a Multiverse of Sieving and Serendipity. Current Anthropology 52 (5): 711–39.

Kohn, Eduardo
    1992   La cultura médica de los Runas de la región amazónica ecuatoriana. Quito: Ediciones Abya-Yala.
    2002a Infidels, Virgins, and the Black-Robed Priest: A Backwoods History of Ecuador's Montaña Region. Ethnohistory 49 (3): 545–82.
    2002b Natural Engagements and Ecological Aesthetics among the Ávila Runa of Amazonian Ecuador. Ph.D. dissertation, University of Wisconsin.
    2005   Runa Realism: Upper Amazonian Attitudes to Nature Knowing. Ethnos 70 (2): 179–96.
    2007   How Dogs Dream: Amazonian Natures and the Politics of Transspecies Engagement. American Ethnologist 34 (1): 3–24.
    2008   Comment on Alexei Yurchak's "Necro-Utopia." Current Anthropology 49 (2): 216–17.

Kull, Kalevi, et al.
    2009   Theses on Biosemiotics: Prolegomena to a Theoretical Biology. Biological Theory 4 (2): 167–73.

Latour, Bruno
    1987   Science in Action. Cambridge, MA: Harvard University Press.
    1993   We Have Never Been Modern. New York: Harvester Wheatsheaf.
    2004   Politics of Nature: How to Bring the Sciences into Democracy. Cambridge, MA: Harvard University Press.
    2005   Reassembling the Social: An Introduction to Actor-Network-Theory. Oxford: Oxford University Press.

Law, John, and Annemarie Mol
    2008   The Actor-Enacted: Cumbrian Sheep in 2001. *In* Material Agency. C. Knappett and M. Lambros, eds. Pp. 57–77. Berlin: Springer.

Lévi-Strauss, Claude
    1966   The Savage Mind. Chicago: University of Chicago Press.
    1969   The Raw and the Cooked: Introduction to a Science of Mythology. Vol. 1. Chicago: University of Chicago Press.

Lévy-Bruhl, Lucien
    1926   How Natives Think. London: Allen & Unwin.

Macdonald, Theodore, Jr.
1979   Processes of Change in Amazonian Ecuador: Quijos Quichua Become Cattlemen. Ph.D. dissertation, University of Illinois, Urbana.

Magnin, Juan
1988 [1740]   Breve descripción de la provincia de Quito, en la América meridional, y de sus misiones... In Noticias auténticas del famoso río Marañon. J. P. Chaumeil, ed. Pp. 463–92. Iquitos, Peru: IIAP-CETA.

Mandelbaum, Allen
1982   The Divine Comedy of Dante Alighieri: Inferno. New York: Bantam Books.

Mannheim, Bruce
1991   The Language of the Inka since the European Invasion. Austin: University of Texas Press.

Margulis, Lynn, and Dorion Sagan
2002   Acquiring Genomes: A Theory of the Origins of Species. New York: Basic Books.

Maroni, Pablo
1988 [1738]   Noticias auténticas del famoso río Marañon. J. P. Chaumeil, ed. Iquitos, Peru: IIAP-CETA.

Marquis, Robert J.
2004   Herbivores Rule. Science 305: 619–21.

Martín, Bartolomé
1989 [1563]   Provanza del Capitan Bartolomé Martín. In La Gobernación de los Quijos. C. Landázuri, ed. Pp. 105–38. Iquitos, Peru: IIAP-CETA.

Mauss, Marcel
1990 [1950]   The Gift: The Form and Reason for Exchange in Archaic Societies. W. D. Halls, trans. New York: Norton.

McFall-Ngai, Margaret, and et al.
2013   Animals in a Bacterial World: A New Imperative for the Life Sciences. Proceedings of the National Academy of Science 110 (9): 3229-36.

McGuire, Tamara L., and Kirk O. Winemiller
1998   Occurrence Patterns, Habitat Associations, and Potential Prey of the River Dolphin, Inia geoffrensis, in the Cinaruco River, Venezuela. Biotropica 30 (4): 625–38.

Mercier, Juan Marcos
1979   Nosotros los Napu-Runas: Napu Runapa Rimay, mitos e historia. Iquitos, Peru: Publicaciones Ceta.

Moran, Emilio F.
1993   Through Amazonian Eyes: The Human Ecology of Amazonian Populations. Iowa City: University of Iowa Press.

Mullin, Molly, and Rebecca Cassidy
2007   Where the Wild Things Are Now: Domestication Reconsidered. Oxford: Berg.

Muratorio, Blanca

    1987   Rucuyaya Alonso y la historia social y económica del Alto Napo, 1850–1950. Quito: Ediciones Abya-Yala.

Nadasdy, Paul

    2007   The Gift in the Animal: The Ontology of Hunting and Human–Animal Sociality. American Ethnologist 34 (1): 25–43.

Nagel, Thomas

    1974   What Is It Like to Be a Bat? Philosophical Review 83 (4): 435–50.

Nietzsche, Friedrich Wilhelm, and R.J. Hollingdale

    1986   Human, All Too Human: A Book for Free Spirits. Cambridge: Cambridge University Press.

Nuckolls, Janis B.

    1996   Sounds Like Life: Sound-Symbolic Grammar, Performance, and Cognition in Pastaza Quechua. New York: Oxford University Press.

    1999   The Case for Sound Symbolism. Annual Review of Anthropology 28: 225–52.

Oakdale, Suzanne

    2002   Creating a Continuity between Self and Other: First-Person Narration in an Amazonian Ritual Context. Ethos 30 (1–2): 158–75.

Oberem, Udo

    1980   Los Quijos: Historia de la transculturación de un grupo indígena en el Oriente ecuatoriano. Otavalo: Instituto Otavaleño de Antropología.

Ochoa, Todd Ramón

    2007   Versions of the Dead: Kalunga, Cuban-Kongo Materiality, and Ethnography. Cultural Anthropology 22 (4): 473–500.

Ordóñez de Cevallos, Pedro

    1989 [1614]   Historia y viaje del mundo. In La Gobernación de los Quijos (1559–1621). Iquitos, Peru: IIAP-CETA.

Orr, Carolyn, and John E. Hudelson

    1971   Cuillurguna: Cuentos de los Quichuas del Oriente ecuatoriano. Quito: Houser.

Orr, Carolyn, and Betsy Wrisley

    1981   Vocabulario quichua del Oriente. Quito: Instituto Lingüístico de Verano.

Orton, James

    1876   The Andes and the Amazon; Or, Across the Continent of South America. New York: Harper and Brothers.

Osculati, Gaetano

    1990   Esplorazione delle regioni equatoriali lungo il Napo ed il Fiume delle Amazzoni: Frammento di un viaggio fatto nell due Americhe negli anni 1846–47–48. Turin, Italy: Il Segnalibro.

Overing, Joanna

    2000   The Efficacy of Laughter: The Ludic Side of Magic within Amazonian Sociality. In The Anthropology of Love and Anger: The Aesthetics of Conviviality in Native Amazonia. J. Overing and A. Passes, eds. Pp. 64–81. London: Routledge.

Overing, Joanna, and Alan Passes, eds.

2000   The Anthropology of Love and Anger: The Aesthetics of Conviviality in Native Amazonia.

Parmentier, Richard J.

1994   Signs in Society: Studies in Semiotic Anthropology. Bloomington: Indiana University Press.

Pedersen, David

2008   Brief Event: The Value of Getting to Value in the Era of "Globalization." Anthropological Theory 8 (1): 57–77.

Peirce, Charles S.

1931   Collected Papers of Charles Sanders Peirce. Cambridge, MA: Harvard University Press.

1992a   The Essential Peirce: Selected Philosophical Writings. Vol. 1. Bloomington: Indiana University Press.

1992b   A Guess at the Riddle. In The Essential Peirce: Selected Philosophical Writings. Vol. 1 (1867–1893). N. Houser and C. Kloesel, eds. Pp. 245–79. Bloomington: Indiana University Press.

1992c   The Law of Mind. In The Essential Peirce: Selected Philosophical Writings. Vol. 1 (1867–1893). N. Houser and C. Kloesel, eds. Pp. 312–33. Bloomington: Indiana University Press.

1992d   Questions Concerning Certain Faculties Claimed for Man. In The Essential Peirce: Selected Philosophical Writings. Vol. 1 (1967–1893). N. Houser and C. Kloesel, eds. Pp. 11–27. Bloomington: Indiana University Press.

1998a   The Essential Peirce: Selected Philosophical Writings. Vol. 2 (1893–1913). Peirce Edition Project, ed. Bloomington: Indiana University Press.

1998b   Of Reasoning in General. In The Essential Peirce: Selected Philosophical Writings. Vol. 2 (1893–1913). Peirce Edition Project, ed. Pp. 11–26. Bloomington: Indiana University Press.

1998c   A Sketch of Logical Critics. In The Essential Peirce: Selected Philosophical Writings. Vol. 2 (1893–1913). Peirce Edition Project, ed. Pp. 451–62. Bloomington: Indiana University Press.

1998d   What Is a Sign? In The Essential Peirce: Selected Philosophical Writings. Vol. 2 (1893–1913). Peirce Edition Project, ed. Pp. 4–10. Bloomington: Indiana University Press.

Pickering, Andrew

1999   The Mangle of Practice: Agency and Emergence in the Sociology of Science. In The Science Studies Reader. M. Biagioli, ed. Pp. 372–93. New York: Routledge.

Porras, Pedro I.

1955   Recuerdos y anécdotas del Obispo Josefino Mons. Jorge Rossi segundo vicario apostólico del Napo. Quito: Editorial Santo Domingo.

1979   The Discovery in Rome of an Anonymous Document on the Quijo Indians of the Upper Napo, Eastern Ecuador. *In* Peasants, Primitives, and Proletariats: The Struggle for Identity in South America. D.L. Browman and R.A. Schwartz, eds. Pp. 13–47. The Hague: Mouton.

Raffles, Hugh
2002   In Amazonia: A Natural History. Princeton, NJ: Princeton University Press.
2010   Insectopedia. New York: Pantheon Books.

Ramírez Dávalos, Gil
1989 [1559]   Información hecha a pedimiento del procurador de la ciudad de Baeça ... *In* La Gobernación de los Quijos. C. Landázuri, ed. Pp. 33–78. Iquitos, Peru: IIAP-CETA.

Rappaport, Roy A.
1999   Ritual and Religion in the Making of Humanity. Cambridge: Cambridge University Press.

Reeve, Mary-Elizabeth
1988   Cauchu Uras: Lowland Quichua Histories of the Amazon Rubber Boom. *In* Rethinking History and Myth: Indigenous South American Perspectives on the Past. J.D. Hill, ed. Pp. 20–34. Urbana: University of Illinois Press.

Requena, Francisco
1903 [1779]   Mapa que comprende todo el distrito de la Audiencia de Quito. Quito: Emilia Ribadeneira.

Riles, Annelise
2000   The Network Inside Out. Ann Arbor: University of Michigan Press.

Rival, Laura
1993   The Growth of Family Trees: Understanding Huaorani Perceptions of the Forest. Man, n.s., 28: 635–52.

Rofel, Lisa
1999   Other Modernities: Gendered Yearnings in China after Socialism. Berkeley: University of California Press.

Rogers, Mark
1995   Images of Power and the Power of Images. Ph.D. dissertation, University of Chicago.

Sahlins, Marshall
1976   The Use and Abuse of Biology: An Anthropological Critique of Sociobiology. Ann Arbor: University of Michigan Press.
1995   How "Natives" Think: About Captain Cook, for Example. Chicago: University of Chicago Press.

Salomon, Frank
2004   The Cord Keepers: Khipus and Cultural Life in a Peruvian Village. Durham, NC: Duke University Press.

Sapir, Edward
  1951 [1929]   A Study in Phonetic Symbolism. *In* Selected Writings of Edward Sapir in Language, Culture, and Personality. D. G. Mandelbaum, ed. Pp. 61–72. Berkeley: University of California Press.
Saussure, Ferdinand de
  1959   Course in General Linguistics. New York: Philosophical Library.
Savage-Rumbaugh, E. Sue
  1986   Ape Language: From Conditioned Response to Symbol. New York: Columbia University Press.
Savolainen, Peter, et al.
  2002   Genetic Evidence for an East Asian Origin of Domestic Dogs. Science 298: 1610–13.
Schaik, Carel P. van, John W. Terborgh, and S. Joseph Wright
  1993   The Phenology of Tropical Forests: Adaptive Significance and Consequences for Primary Consumers. Annual Review of Ecology and Systematics 24: 353–77.
Schwartz, Marion
  1997   A History of Dogs in the Early Americas. New Haven, CT: Yale University Press.
Silverman, Kaja
  2009   Flesh of My Flesh. Stanford, CA: Stanford University Press.
Silverstein, Michael
  1995   Shifters, Linguistic Categories, and Cultural Description. *In* Language, Culture, and Society. B. G. Blount, ed. Pp. 187–221. Prospect Heights, IL: Waveland Press.
Simson, Alfred
  1878   Notes on the Záparos. Journal of the Anthropological Institute of Great Britain and Ireland 7: 502–10.
  1880   Notes on the Jívaros and Canelos Indians. Journal of the Anthropological Institute of Great Britain and Ireland 9: 385–94.
Singh, Bhrigupati
  2012   The Headless Horseman of Central India: Sovereignty at Varying Thresholds of Life. Cultural Anthropology 27 (2): 383–407.
Slater, Candace
  2002   Entangled Edens: Visions of the Amazon. Berkeley: University of California Press.
Smuts, Barbara
  2001   Encounters with Animal Minds. Journal of Consciousness Studies 8 (5–7): 293–309.
Stevenson, Lisa
  2012   The Psychic Life of Biopolitics: Survival, Cooperation, and Inuit Community. American Ethnologist 39 (3): 592–613.

Stoller, Paul

1997   Sensuous Scholarship. Philadelphia: University of Pennsylvania Press.

Strathern, Marilyn

1980   No Nature: No Culture: The Hagen Case. In Nature, Culture, and Gender. C. MacCormack and M. Strathern, eds. Pp. 174–222. Cambridge: University of Cambridge Press.

1988   The Gender of the Gift: Problems with Women and Problems with Society in Melanesia. Berkeley: University of California Press.

1995   The Relation: Issues in Complexity and Scale. Vol. 6. Cambridge: Prickly Pear Press.

2004 [1991]   Partial Connections. Walnut Creek, CA: AltaMira Press.

Suzuki, Shunryu

2001   Zen Mind, Beginner's Mind. New York: Weatherhill.

Taussig, Michael

1987   Shamanism, Colonialism, and the Wild Man: A Study in Terror and Healing. Chicago: University of Chicago Press.

Taylor, Anne Christine

1993   Remembering to Forget: Identity, Mourning and Memory among the Jivaro. Man, n.s., 28: 653–78.

1996   The Soul's Body and Its States: An Amazonian Perspective on the Nature of Being Human. Journal of the Royal Anthropological Institute, n.s., 2: 201–15.

1999   The Western Margins of Amazonia From the Early Sixteenth to the Early Nineteenth Century. In The Cambridge History of the Native Peoples of the Americas. F. Salomon and S.B. Schwartz, eds. Pp. 188–256. Cambridge: Cambridge University Press.

Tedlock, Barbara

1992   Dreaming and Dream Research. In Dreaming: Anthropological and Psychological Interpretations. B. Tedlock, ed. Pp. 1–30. Santa Fe, NM: School of American Research Press.

Terborgh, John

1990   Mixed Flocks and Polyspecific Associations: Costs and Benefits of Mixed Groups to Birds and Monkeys. American Journal of Primatology 21 (2): 87–100.

Tsing, Anna Lowenhaupt

2012   On Nonscalability: The Living World Is Not Amenable to Precision-Nested Scales. Common Knowledge 18 (3): 505–24.

Turner, Terence

1988   Ethno-Ethnohistory: Myth and History in Native South American Representations of Contact with Western Society. In Rethinking History and Myth: Indigenous South American Perspectives on the Past. J.D. Hill, ed. Pp. 235–81. Urbana: University of Illinois Press.

2007   The Social Skin. In Beyond the Body Proper. M. Lock and J. Farquhar, eds. Pp. 83–103. Durham, NC: Duke University Press.

Tylor, Edward B.
   1871   Primitive Culture: Researches into the Development of Mythology, Philoso-
          phy, Religion, Art, and Custom. London: J. Murray.
Urban, Greg
   1991   A Discourse-Centered Approach to Culture: Native South American Myths
          and Rituals. Austin: University of Texas Press.
Uzendoski, Michael
   2005   The Napo Runa of Amazonian Ecuador. Urbana: University of Illinois
          Press.
Venkatesan, Soumhya, et al.
   2010   Debate: Ontology Is Just Another Word for Culture. Critique of Anthropol-
          ogy 30 (2): 152–200.
Vilaça, Aparecida
   2007   Cultural Change as Body Metamorphosis. In Time and Memory in Indige-
          nous Amazonia: Anthropological Perspectives. C. Fausto and M. Hecken-
          berger, eds. Pp. 169–93. Gainesville: University Press of Florida.
   2010   Strange Enemies: Indigenous Agency and Scenes of Encounters in Amazo-
          nia. Durham, NC: Duke University Press.
Viveiros de Castro, Eduardo
   1998   Cosmological Deixis and Amerindian Perspectivism. Journal of the Royal
          Anthropological Institute, n.s., 4: 469–88.
   2009   Métaphysiques cannibales: Lignes d'anthropologie post-structurale. Paris:
          Presses universitaires de France.
von Uexküll, Jakob
   1982   The Theory of Meaning. Semiotica 42 (1): 25–82.
Wavrin, Marquis Robert de
   1927   Investigaciones etnográficas: Leyendas tradicionales de los Indios del Ori-
          ente ecuatoriano. Boletín de la Biblioteca Nacional, n.s., 12: 325–37.
Weber, Max
   1948a Religious Rejections of the World and Their Directions. In From Max
          Weber: Essays in Sociology. H. H. Gerth and C. W. Mills, eds. Pp. 323–59.
          Oxon: Routledge.
   1948b Science as a Vocation. In From Max Weber: Essays in Sociology. H. H.
          Gerth and C. W. Mills, eds. Pp. 129–56. Oxon: Routledge.
Weismantel, Mary J.
   2001   Cholas and Pishtacos: Stories of Race and Sex in the Andes. Chicago: Uni-
          versity of Chicago Press.
White, Richard
   1991   The Middle Ground: Indians, Empires, and Republics in the Great Lakes
          Region, 1650–1815. Cambridge: Cambridge University Press.
Whitten, Norman E.
   1976   Sacha Runa: Ethnicity and Adaptation of Ecuadorian Jungle Quichua.
          Urbana: University of Illinois Press.

1985   Sicuanga Runa: The Other Side of Development in Amazonian Ecuador. Urbana: University of Illinois Press.

Willerslev, Rane

2007   Soul Hunters: Hunting, Animism, and Personhood among the Siberian Yukaghirs. Berkeley: University of California Press.

Wills, Christopher, et al.

1997   Strong Density and Diversity-Related Effects Help to Maintain Tree Species Diversity in a Neotropical Forest. Proceedings of the National Academy of Science, no. 94: 1252–57.

Yurchak, Alexei

2006   Everything Was Forever, Until It Was No More: The Last Soviet Generation. Princeton, NJ: Princeton University Press.

2008   Necro-Utopia. Current Anthropology 49 (2): 199–224.

INDEX